THE PULSE

OF THE BAY 2019

Pollutant Pathways

A REPORT OF THE REGIONAL MONITORING PROGRAM
FOR WATER QUALITY IN SAN FRANCISCO BAY

SUGGESTED CITATION: SFEI. 2019. The Pulse of the Bay: Pollutant Pathways. SFEI Contribution #954. San Francisco Estuary Institute, Richmond, CA.

OVERVIEW

Pollutants make their way to San Francisco Bay from near and far. Some pollutants come from the other side of the world, such as the mercury that is emitted into the atmosphere by coal-fired power plants in Asia, transported across the Pacific Ocean by the wind, and deposited on the surface of the Bay and in the Bay watershed. Some pollutants come from the far corners of the Bay's watershed, which encompasses 40% of the land surface of California. Mercury used in historic gold mining regions in the Sierra Nevada is still flowing downstream into Central Valley rivers, through the Delta, and into the Bay.

Most pollutants, however, originate from activities closer to home in the small, local watersheds that surround the Bay. Rain that falls onto yards, roofs, parking lots, streets, farms, pastures, and other surfaces in the watersheds picks up pollutants and carries them to the Bay through an intricate network of thousands of miles of storm drain pipes, culverts, and creeks. Most of this stormwater flows to the Bay without any form of pollutant removal.

Another parallel and intricate network of thousands of miles of pipes carries wastewater from homes, business, and industries through sewage collection systems to municipal wastewater treatment plants. These treatment plants are highly effective at removing most pollutants from the wastewater stream, but some are not effectively treated and pass through to the Bay.

Some industrial operations, such as the Bay Area's petroleum refineries, are large enough to warrant having their own wastewater treatment facilities that remove pollutants so the effluent can be safely discharged into the Bay, reused in plant operations, or applied to land.

Disposal of dredged material at designated sites within the Bay is another activity that introduces pollutants to specific areas in the Bay, but in this case the pollutants are simply transferred from one part of the Bay to another. Much of the pollutant mass in dredged material is also removed from the Bay via disposal in the ocean or beneficial reuse in wetlands and upland sites.

View from Angel Island (Shira Bezalel, SFEI)

This edition of the Pulse of the Bay features articles on the four major pollutant pathways that are the primary focus of Bay water quality managers: municipal wastewater, industrial wastewater, stormwater, and dredging and dredged material disposal. Management of these pathways will be the key to further improving and protecting Bay water quality. For each pathway, the articles provide an introduction summarizing origins of the pollutants, steps that are taken to remove pollutants, regulations that are the basis for management, findings from recent studies, and a discussion of future directions and challenges.

Billions of dollars are spent every year to remove pollutants from these pathways and protect Bay water quality. The estimated annual cost in 2015 of operation, maintenance, and capital investment for Bay Area sewage collection and treatment systems alone was $4.3 billion. These expenditures are likely to grow as Bay Area municipalities and industries

update and enhance their aging infrastructure to further improve Bay water quality and meet the challenges of population growth, climate change, and sea level rise. As just one recent example of this type of investment infrastructure, Silicon Valley Clean Water (which serves the cities of Belmont, San Carlos, and Redwood City) embarked this summer on a $495 million project to replace and rehabilitate their sewage conveyance system.

Bay Area stormwater management agencies are also facing the challenge of storm drainage infrastructure reaching the end of its lifespan, as much of it was constructed over 50 years ago. Replacing this gray infrastructure with green stormwater infrastructure (or GSI – features such as rain gardens, permeable pavements, and green streets) is a major component of plans to meet stormwater load reduction goals for PCBs and mercury, as well as controlling other pollutants and providing other environmental and social benefits. A

(CONTINUED ON PAGE vi)

Pollutant Sources and Pathways to the Bay

Activities in our homes, businesses, and industrial facilities send polluted water into the sewage collection system. **MUNICIPAL WASTEWATER** treatment plants are highly effective at removing much of the pollution. However, the common forms of treatment only partially remove some pollutants, including nutrients and many contaminants of emerging concern.

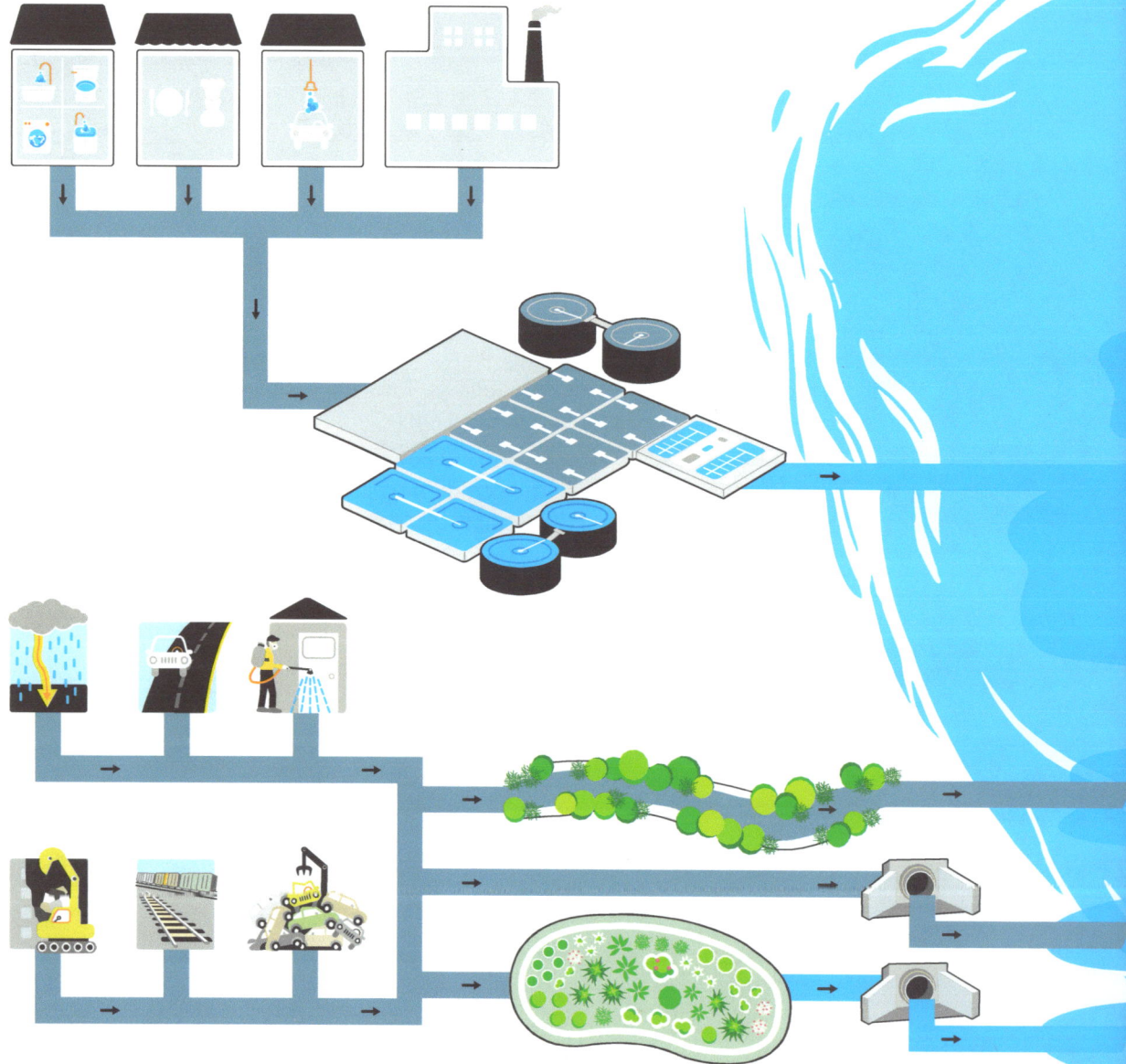

Activities such as vehicle use, pesticide application, building demolition, and illegal dumping pollute rainwater that flows off land surfaces to the Bay. Deposition of pollutants from the atmosphere to the watershed and the Bay is another important pathway. Most **STORMWATER** is untreated, but increasing amounts are being filtered by green stormwater infrastructure. Urban stormwater has a high potential and need for load reduction for some of the Bay pollutants of greatest concern.

The Sacramento and San Joaquin rivers together drain a vast area – 37% of the state of California – and enter the Bay after flowing through the Delta. This **DELTA OUTFLOW** contributes 89% of the fresh water that enters the Bay, as well as large loads of many pollutants of concern (mercury, PCBs, selenium, nitrogen), but generally in a relatively dilute form.

Petroleum refineries account for most of the **INDUSTRIAL WASTEWATER** discharged to the Bay. Refinery wastewater treatment plants employ the same basic elements as in treatment of municipal wastewater, but add other processes to remove oil, hydrocarbons, and selenium. Refinery effluent contributes more than 1% of the total regional loading of only one pollutant: selenium.

DREDGING AND DISPOSAL OF DREDGED MATERIAL moves pollutants from one part of the Bay to another, or removes them from the Bay entirely. Dredging and disposal at in-Bay disposal sites uncover and remobilize sediment-bound contaminants, such as PCBs and mercury. Disposal of dredged sediment in the ocean or at upland sites, or re-using it in wetland restoration, removes pollutants from the Bay.

Illustration by Linda Wanczyk (lindawanczyk.com)

major barrier to implementing GSI plans, however, is a lack of adequate funding. Most Bay Area municipalities have limited or no dedicated funding for stormwater management, including for GSI implementation.

Meeting the load reduction goals for stormwater is one of the biggest hurdles to improving Bay water quality. In addition to integrating GSI into the urban landscape, abatement of known sources in the watersheds, and management of PCB-containing materials during building demolition will be key actions.

Managing nitrogen loads from municipal wastewater treatment plants to keep the Bay free of problematic algae blooms is another priority. A nitrogen load increase of 5% over the last six years is thought to have been driven by growing population, economic conditions that attract workers to the Bay Area, and increasing importation of food waste and agricultural waste to wastewater treatment plants for energy production. Given the projected continued increase in Bay Area population, the recent trend data indicate that loads are likely to rise by 1-2% per year unless nitrogen removal is further enhanced. Many municipalities plan to improve their facilities to reduce their nitrogen loads by 2024.

Sea level rise and the more intense droughts and floods that are anticipated due to climate change will impact all of the pollutant pathways. Most municipal and industrial wastewater treatment facilities are in low-lying areas on the edge of the Bay and are vulnerable to sea level rise. Sea level rise may also imperil low-lying storm drain infrastructure and expose contaminated shoreline areas to the forces of tides and waves. Dredged sediment has become a precious resource that is needed to restore wetlands and pursue nature-based solutions to sea level rise.

Larger swings in the volumes of wastewater and stormwater flow that have to be managed will increase infrastructure costs. Higher flows from more intense storms will drive a need for increased capacity in wastewater and stormwater management systems. These high flows will lead to greater pollutant loads from stormwater, due to increased mobilization from the watershed – as recently observed in mercury loads from the Guadalupe River (**page 84**) – and the fact that GSI is not designed to handle intense storms. Reductions in indoor water use during droughts can damage wastewater infrastructure and reduce the effectiveness of treatment processes.

Droughts also highlight the increasing focus on wastewater and stormwater as valuable resources. Volumes of flow to the Bay are expected to decline as these streams are increasingly recycled and reused. Municipal wastewater will be regarded less as a waste to be discarded, and more and more as a valuable source of water, soil amendments, and energy.

Contaminants of emerging concern pose another challenge for management of municipal wastewater and stormwater. Based on findings from recent RMP studies, five CECs have been added to the moderate concern category in the past two years (fluorinated stain-repellants, the insecticide imidacloprid, bisphenol plastic additives, organophosphate ester flame retardants and plastic additives, and microplastics) (**page 76**). These pollutants are generally not effectively removed by municipal wastewater treatment plants, and recent studies also point to stormwater as a major pathway.

As the Bay Area faces all of these changes and challenges, the need for robust regional monitoring remains in order to ensure that the effort and resources invested in managing pollutant pathways are effective in protecting and improving the health of San Francisco Bay. §

CONTENTS

▲ **Double-crested Cormorant on pilings** (Shira Bezalel, SFEI)

FEATURE
ARTICLES

Looking west from Point Richmond (Shira Bezalel, SFEI)

MUNICIPAL WASTEWATER

Jay Davis
San Francisco Estuary Institute

Lorien Fono and **David Williams**
Bay Area Clean Water Agencies

Bill Johnson and **Robert Schlipf**
*San Francisco Bay
Regional Water Quality Control Board*

Tom Hall
EOA, Inc.

HIGHLIGHTS

▶ Pollutants in municipal wastewater enter sewer systems from homes, businesses, and industrial facilities, and are largely removed from the waste stream by wastewater treatment plants

▶ Municipal wastewater is highly regulated, perhaps the most closely regulated pollutant pathway

▶ Municipal wastewater treatment improvements have occurred in phases, with primary treatment (solids removal) in the 1950s and 1960s, secondary treatment (organics removal) in the 1970s and 1980s, a focus on pretreatment and pollution prevention for toxics control in the 1980s and 1990s, and now consideration of major improvements to reduce nutrient discharges and shift toward resource recovery

▶ While municipal wastewater is no longer the most important pathway for many pollutants entering San Francisco Bay, it is the primary pathway for nutrients and many contaminants of emerging concern

▶ Municipal wastewater infrastructure is aging, and infrastructure planning needs to address nutrients, climate change, and other evolving issues

▶ Municipal wastewater is increasingly regarded as a valuable resource: a source of water, soil amendments, and energy

◀ **The East Bay Municipal Utility District main wastewater treatment plant** (Alamy)

Municipal Wastewater 101

When the 7.8 million people living in the Bay Area flush their toilets or send flows down the drain via sinks, bathtubs, showers, and washing machines, the dirty water begins a journey through a vast regional network of thousands of miles of sewer pipes, past pump stations, to a wastewater treatment plant that removes most of the pollutants before discharging treated municipal wastewater effluent to the Bay or a Bay tributary (**Figure 1**). Thousands of commercial and industrial facilities also discharge to Bay Area sewage collection and treatment systems. The San José-Santa Clara Regional Wastewater Facility, for example, the largest in the Bay Area, serves approximately 17,000 commercial and industrial connections in addition to 1.5 million people.

The sewage collection and treatment systems of the Bay Area represent a massive investment in protecting environmental quality in our neighborhoods and water quality in the Bay and its tributaries. Billions of dollars have been spent to build these systems, which include 45 wastewater treatment plants (**Figure 2**), and billions more are spent every year to operate and maintain them. In 2015 the Bay Area Clean Water Agencies (an association that represents most of the municipalities and special districts that provide sewer services in the Bay Area) estimated the annual overall budget for their wastewater agencies to be $4.3 billion: approximately $1.8 billion for capital investment and $2.5 billion for operations and maintenance.

Although wastewater treatment plants are sometimes misconstrued as pollution sources, they are really part of the sewage collection

Figure 1. An example sewage collection system. (Top) The City of Oakland has approximately 900 miles of wastewater collection system sewers and seven pump stations that serve a population of about 400,000 people. (Bottom) Oakland and seven other East Bay communities convey wastewater to interceptors that lead to the East Bay Municipal Utility District wastewater treatment plant (EBMUD 2016).

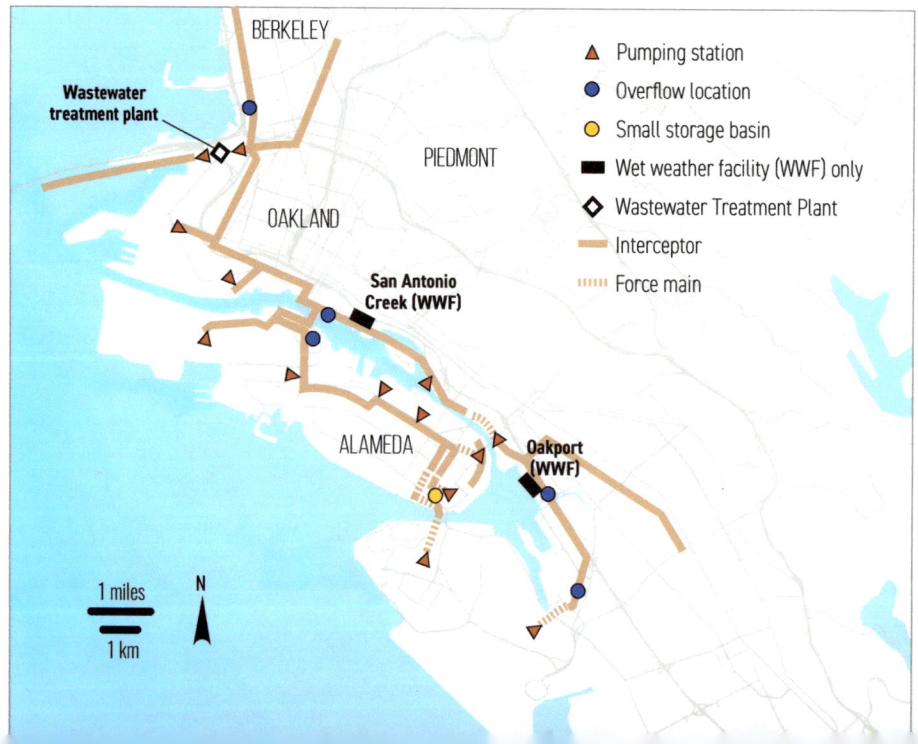

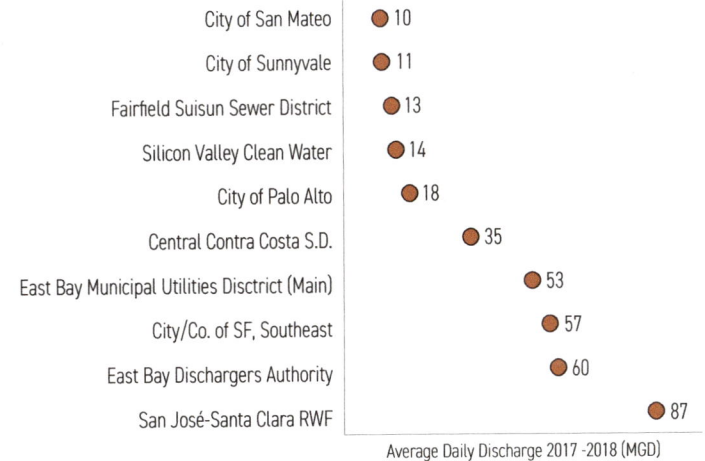

Figure 2. Locations of the 38 municipal wastewater outfalls in the Bay Area.

The average total volume of flow from these plants in 2017-2018 was 434 million gallons per day. Most of these plants provide secondary treatment (**page 10**); 11 have advanced secondary treatment.

FOOTNOTE: Adapted from SFBRWQCB (2019). Flow data from BACWA (2018).

- ● Municipal wastewater outfall
- ◌ Ten largest WWTP contributions by average daily discharge (2017-2018)
- —— Water Board boundary

5 miles
5 km

N

Municipal Dischargers[1]:

1. City of American Canyon* (1.4)
2. City of Benicia (2.0)
3. City of Burlingame (2.8)
   City of Millbrae (1.5)
   City/Co. of SF, Int'l Airport (1.2)
   So. SF/San Bruno WQCP (7.6)
4. City of Calistoga* (0.3)
5. Central Contra Costa S.D. (35.4)
6. Central Marin Sanitation A.G. (9.3)
7. Crockett Community Services District, Port Costa (0.02)
8. Delta Diablo (9.6)
9. East Bay Dischargers Authority (59.7):
   -City of Hayward
   -Oro Loma S.D.
   -Castro Valley S.D.
   -City of San Leandro
   -Union S.D.
   Livermore-Amador Valley WMA:
   -Dublin/San Ramon S.D.
   -City of Livermore
10. EBMUD (Main wastewater plant) (52.5)
11. EBMUD (Pt. Isabel)
12. EBMUD (San Antonio Creek)
13. EBMUD (Oakport)

15. Fairfield Suisun Sewer District* (13.4)
16. Las Gallinas Valley S.D. (1.4)
17. Marin Co. S.D. No. 5 (Tiburon) (0.62)
18. Marin Co. S.D. No. 5 (Paradise) (0.01)
19. Mountain View S.D.* (1.3)
20. Napa S.D. (4.6)
21. Novato S.D. (3.0)
22. City of Palo Alto* (18.4)
23. City of Petaluma (3.2)
24. Cities of Pinole & Hercules Rodeo S.D. (2.5)
25. City of Saint Helena* (0.2)
26. City/Co. of SF, Southeast (57.4)
27. San José-Santa Clara RWF* (87)
28. City of San Mateo (10.4)
29. Sausalito-Marin City S.D. (1.2)
31. Sewerage Agency of S. Marin (2.3)
32. Sonoma Valley County S.D.* (0.0)
33. Silicon Valley Clean Water* (14.0)
34. City of Sunnyvale* (10.6)
35. U.S. Navy Treasure Island (0.3)
36. Vallejo Sanitation & Flood Control (9.2)
37. West County Agency (9.8):
   -City of Richmond
   -West County Wastewater District
38. Town of Yountville* (0.1)

[1]Average daily discharge in MGD for 2017-2018 indicated in parentheses
* WWTP performs advanced secondary treatment

City of San Mateo ● 10
City of Sunnyvale ● 11
Fairfield Suisun Sewer District ● 13
Silicon Valley Clean Water ● 14
City of Palo Alto ● 18
Central Contra Costa S.D. ● 35
East Bay Municipal Utilities Disctrict (Main) ● 53
City/Co. of SF, Southeast ● 57
East Bay Dischargers Authority ● 60
San José-Santa Clara RWF ● 87

Average Daily Discharge 2017 -2018 (MGD)

Ten largest flows for 2017-2018

and treatment system *pathway*. The real sources of pollution in municipal wastewater are the activities in our homes, businesses, and industries that send polluted water down the drain and into the sewage system (**pages 8-9**). Human waste is of course a major source of organic matter, bacteria, and nutrients, but also of pharmaceuticals and other chemicals that pass through our bodies, as well as many chemical metabolites. Other major categories of chemicals that are used in homes and businesses that pose concerns for water quality include cleaning products, pesticides, stain repellants, personal care products, flame retardants, and plastics and plastic additives. Industrial facilities also discharge wastes to municipal sewer systems and can be sources of pollutants such as chemicals used in manufacturing goods and products.

Wastewater treatment plants are highly effective at removing solids, organic matter, and bacteria from the waste stream. They often remove toxic pollutants, too, if the toxic pollutants are biodegradable or adhere to solids. Municipal wastewater treatment plants in the Bay Area all provide a minimum of secondary treatment, which typically includes screening, skimming, settling, and biological treatment (**page 10**). Eleven plants that discharge to the Bay all provide advanced secondary treatment, which can include conversion of ammonia to nitrate or removal of additional solids, often with filtration. Removing additional solids improves removal of pollutants that adhere to particles, including mercury and polychlorinated biphenyls (PCBs). There are some plants that provide additional advanced treatment and also denitrify at various levels, which removes nitrogen from wastewater.

Municipal wastewater is increasingly viewed less as a waste and more as a valuable resource. Water is often a scarce commodity in California, as recently highlighted by the five-year drought from 2012-2016. Water scarcity is expected to increase over time in the Bay Area and across the country as a result of drought, growing water demand, and other stressors. Municipal wastewater can be treated and reused, and serve as a drought-proof, dependable, and local water supply.

Many municipalities "recycle" filtered treated wastewater – using it for irrigating landscaping, parks, and golf courses; for dust control at construction sites; for industrial supply; and as a supply for toilets and fountains. Water recycling is becoming more and more common. For example, it is projected that

municipalities discharging to the Lower South Bay will be recycling more than 50% of their wastewater by 2035.

With additional treatment, using techniques such as microfiltration, reverse osmosis, advanced oxidation, and ultraviolet disinfection, municipal wastewater can produce highly purified (near-distilled quality) water that is suitable for a wide variety of uses, including drinking water for humans ("potable reuse"). While these additional treatment technologies produce highly purified water, they also produce a concentrated waste stream ("concentrate") that must also be managed and disposed of. There are different types of potable reuse based on whether the recycled water passes through an environmental buffer (e.g., groundwater aquifer, lake, or river) before the water is again treated and used as drinking water. "Groundwater augmentation" is the use of recycled water for replenishment of a groundwater basin or an aquifer that has been designated as drinking water. "Raw water augmentation" is the placement of recycled water into a system of pipelines or aqueducts that deliver the water to a drinking water treatment plant. "Reservoir water augmentation" is the placement of recycled water into a raw surface water reservoir used as a source of drinking water, or into a constructed system conveying water to such a reservoir. Another type of potable reuse requires a higher level of initial treatment followed by storage and use, but without the environmental buffer. "Treated drinking water augmentation" is the placement of highly treated recycled water directly into a drinking water distribution system. Across the US, an increasing number of drinking water systems rely on some form of potable reuse.

A Bay Area leader in performing this highest level of treatment is the Silicon Valley Advanced Water Purification Center, the largest advanced water purification plant in Northern California. The Purification Center uses microfiltration, reverse osmosis, and ultraviolet light to further process secondary treated water from the San José-Santa Clara Regional Wastewater Facility and generates up to eight million gallons a day of purified water that meets California's primary drinking water standards. The $72 million Purification Center is a partnership between Valley Water and the City of San José, and began operation in 2014. Currently, the purified water produced at the Purification Center is blended with the existing recycled water supply produced at the Regional Wastewater Facility to enhance water quality and expand recycled water use. In the future, highly purified water from the

Center may be used for a variety of purposes, including expanding Silicon Valley's drinking water supply.

The solids that treatment plants remove from municipal wastewater (**page 10**) are also a valuable resource. Solids produced from the wastewater treatment process can be converted to a soil amendment by heating them to a high temperature in "digesters" to reduce the disease-causing organisms and break down the organic matter. The material from the digesters is called "biosolids." Biosolids are regularly monitored to ensure that they meet or surpass quality and safety standards established by the US Environmental Protection Agency (USEPA), State of California, and local governments. The East Bay Municipal Utility District (EBMUD) is an example of a Bay Area facility with an active biosolids program. Currently, all biosolids produced by EBMUD (60,000 to 70,000 wet tons per year) are beneficially reused as a soil amendment at nearby non-food crop farms and as alternative daily cover at local landfills.

In spite of the general effectiveness of municipal wastewater treatment plants, and their successful elimination of serious past water quality concerns in the Bay, some pollutants are only partially removed by the plants, and municipal wastewater remains an important pathway by which these pollutants enter San Francisco Bay. Pollutants that do not readily settle out of the water and are resistant to microbial digestion, such as some pharmaceuticals and pesticides, pass through the plant with limited removal. In addition, municipal wastewater treatment plants with secondary treatment only remove around 25% of the nitrogen that flows into the plants. As part of the Nutrient Watershed Permit (discussed further below), several Bay Area facilities plan to add to or enhance their nitrification and denitrification capabilities to reduce their nitrogen loads by 2024.

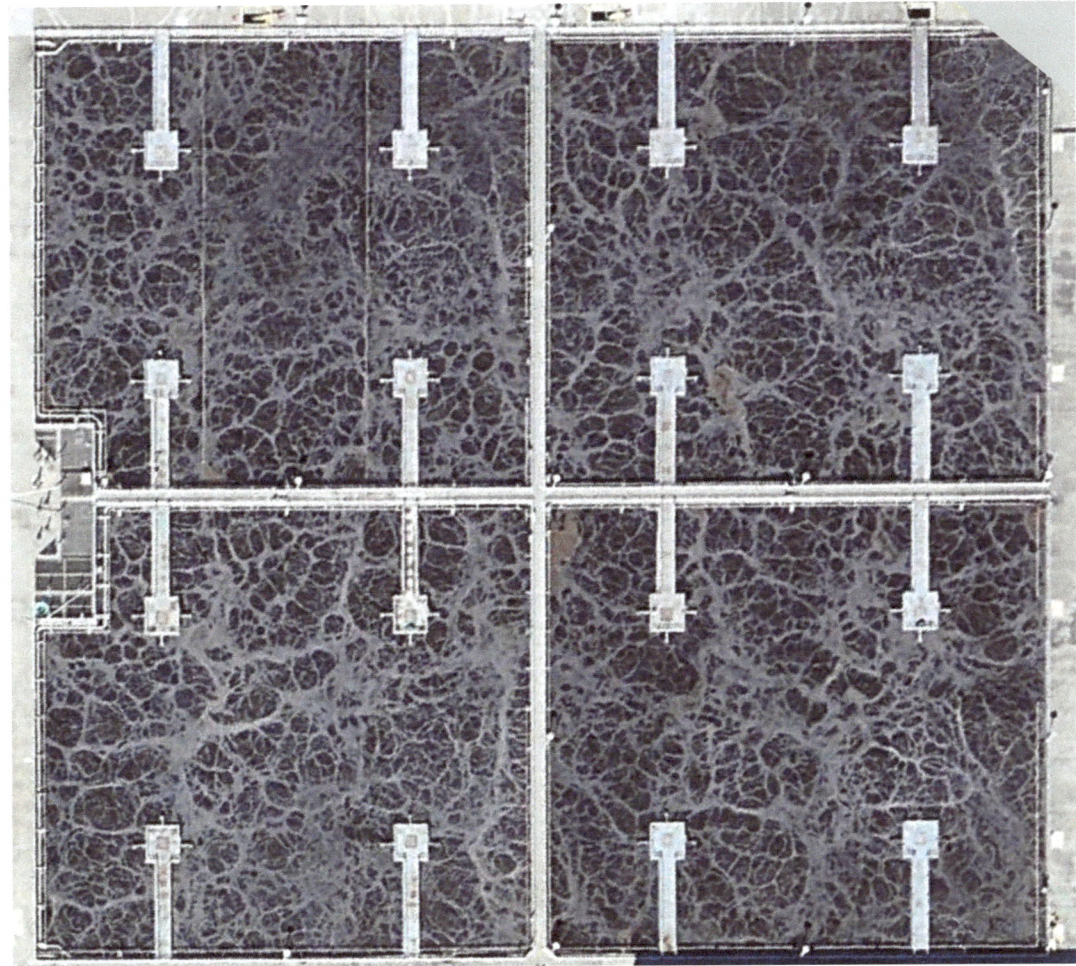

▲ **Aeration basins at the Palo Alto Regional Water Quality Control Plant**
(Google Earth)

Municipal Wastewater: Sources, Pathways, Loading

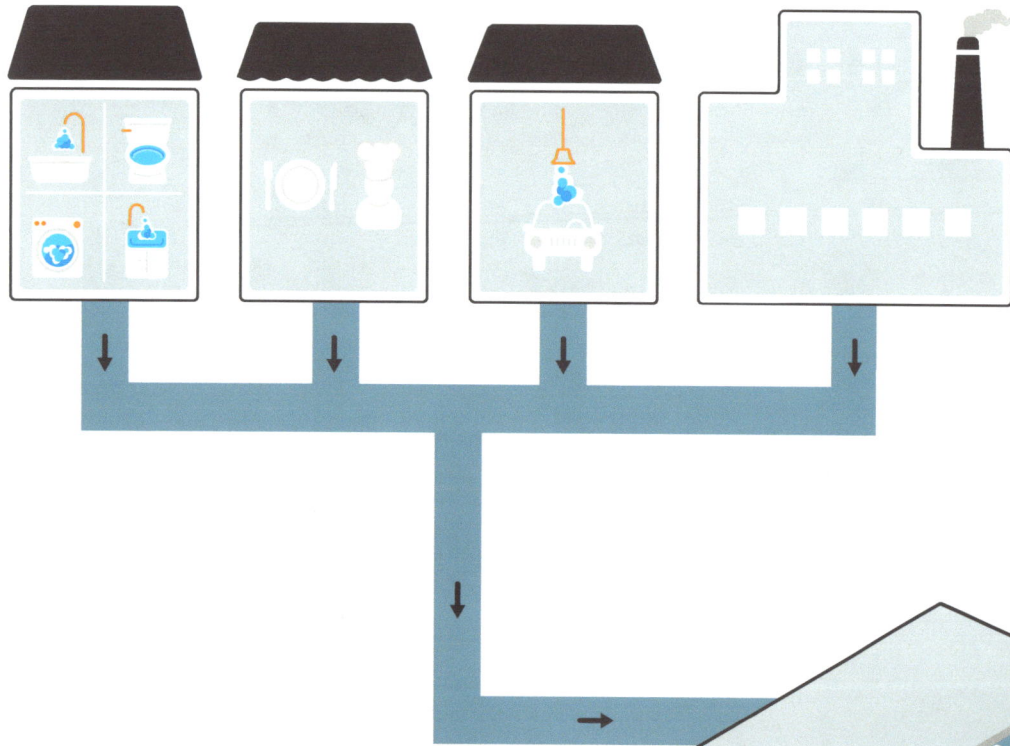

The **SOURCES** of pollution in municipal wastewater are the activities in our homes, businesses, and industrial facilities that send polluted water down the drain and into the sewage system. Human waste is a major source of organic matter, bacteria, nutrients, and pharmaceuticals. Homes and businesses are also sources of cleaning products, pesticides, stain repellants, personal care products, flame retardants, plastics, and plastic additives. Industrial facilities also discharge to municipal sewer systems, and are sources of other pollutants.

Municipal wastewater treatment plants are **PATHWAYS** by which some of these pollutants enter the Bay. Wastewater treatment plants are highly effective at removing solids, organic matter, and bacteria from the waste stream. They often remove toxic pollutants, too, if the toxic pollutants are biodegradable or adhere to solids. More advanced forms of treatment can also remove nutrients and toxic pollutants.

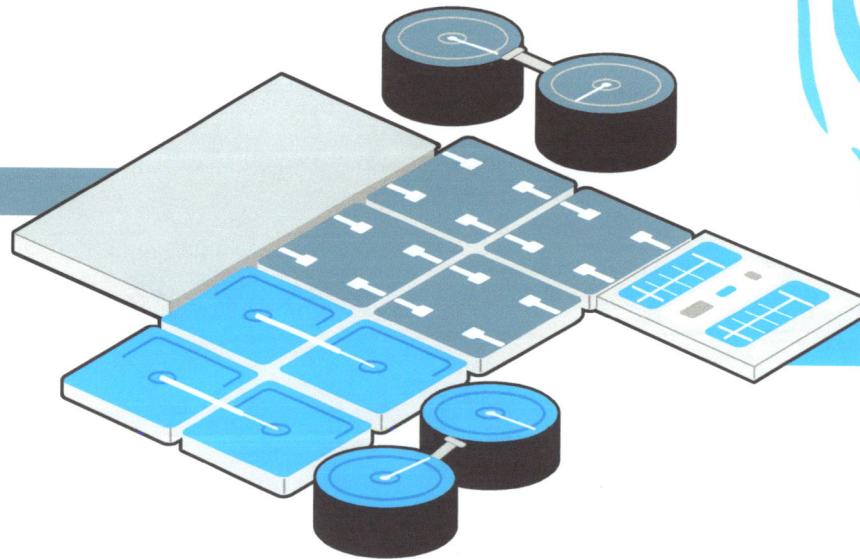

A significant percentage of the **LOADING** of some pollutants of concern in the Bay is attributable to municipal wastewater. The common forms of treatment in the Bay Area only partially remove these pollutants, which include nutrients and many contaminants of emerging concern, such as pesticides, fluorinated stain repellants, surfactants from detergents and cleaning products, plastic additives such as bisphenols, flame retardants, and microplastics.

Illustration by Linda Wanczyk (lindawanczyk.com)

Municipal Wastewater Treatment

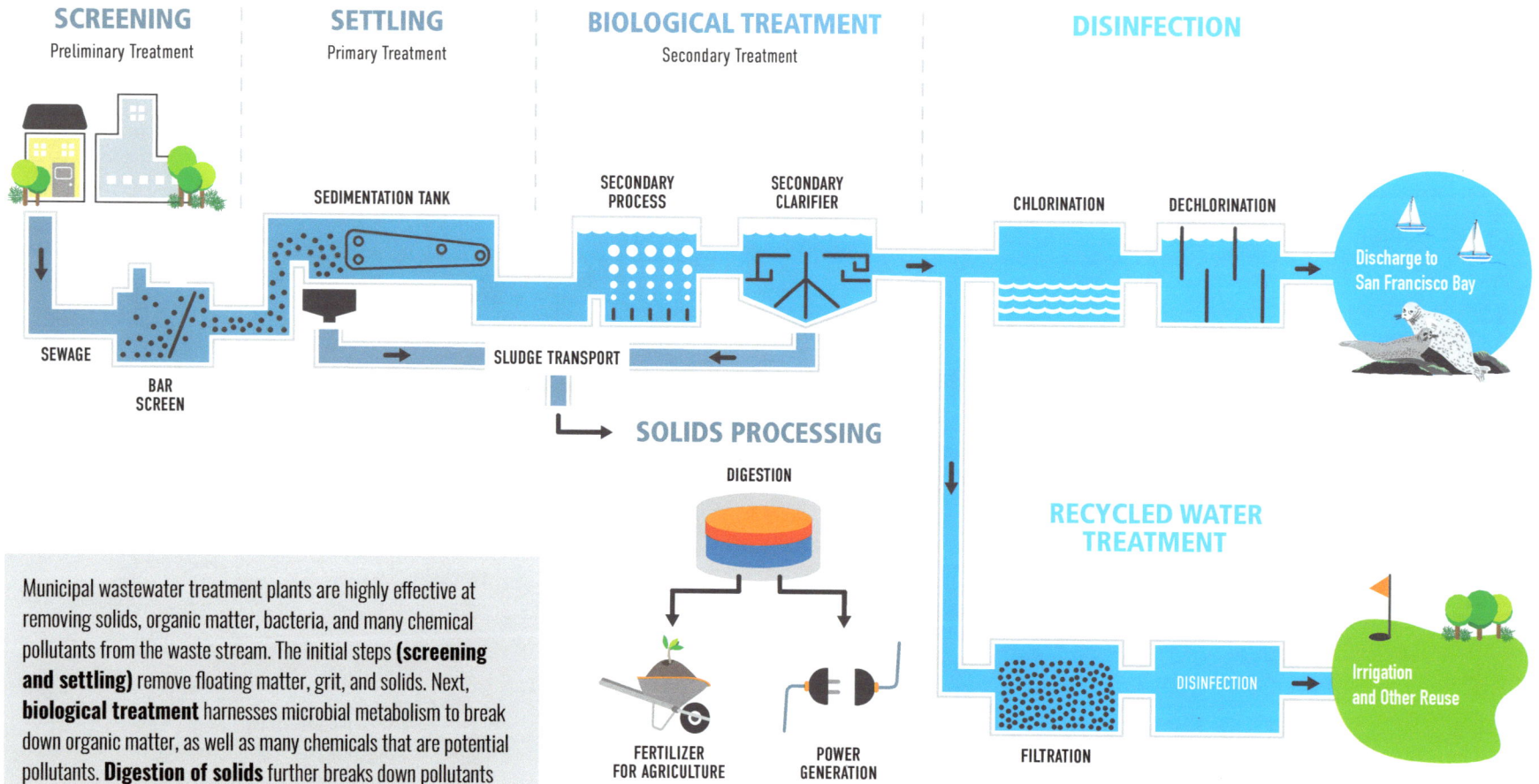

SCREENING
Preliminary Treatment

SETTLING
Primary Treatment

BIOLOGICAL TREATMENT
Secondary Treatment

DISINFECTION

SEDIMENTATION TANK

SECONDARY PROCESS

SECONDARY CLARIFIER

CHLORINATION

DECHLORINATION

Discharge to San Francisco Bay

SEWAGE

BAR SCREEN

SLUDGE TRANSPORT

SOLIDS PROCESSING

DIGESTION

FERTILIZER FOR AGRICULTURE

POWER GENERATION

RECYCLED WATER TREATMENT

FILTRATION

DISINFECTION

Irrigation and Other Reuse

Municipal wastewater treatment plants are highly effective at removing solids, organic matter, bacteria, and many chemical pollutants from the waste stream. The initial steps **(screening and settling)** remove floating matter, grit, and solids. Next, **biological treatment** harnesses microbial metabolism to break down organic matter, as well as many chemicals that are potential pollutants. **Digestion of solids** further breaks down pollutants that adhere to solids. Pollutants that do not readily settle out of the water and are resistant to microbial digestion pass through the plant to the Bay. Pollutants that do adhere to solids but are very persistent may still be present in the fertilizer produced from solids processing. Disinfection by either chlorine or ultraviolet light inactivates bacteria and viruses (pathogens) to levels safe to discharge to the Bay or for recycled water applications.

Illustration by Linda Wanczyk (lindawanczyk.com)
Adapted from an illustration by the Palo Alto Regional Water Quality Control Plant.

Regulatory Framework

The NPDES Program Governs Wastewater Nationwide

Discharges of municipal wastewater into San Francisco Bay and its tributaries have a long history of careful regulation. The National Pollutant Discharge Elimination System (NPDES) Program provides the framework by which this regulation occurs in the Bay Area and across the nation. The NPDES Program is one of the most successful environmental programs ever implemented. Since the federal Clean Water Act created this program in 1972, pollutant loading to the Bay from municipal wastewater has been dramatically reduced, leading to profound improvements in Bay water quality. The San Francisco Bay Regional Water Quality Control Board (Water Board) is the agency with the primary responsibility for regulating Bay water quality. Pursuant to the Clean Water Act, USEPA delegated most aspects of the NPDES Program to the Water Board. Over the last five decades, the Water Board has used the NPDES Program as a framework for implementing highly innovative and adaptive enhancements to the management of municipal wastewater discharges and other pollutant pathways to the Bay.

Wastewater generators must obtain a NPDES permit to discharge their wastewater into a water body. NPDES permits contain specific requirements that limit the pollutants in discharges. They also require dischargers to monitor their wastewater to ensure that it meets all requirements. Wastewater dischargers must maintain their treatment facilities, and treatment plant operators must be certified. The Water Board regularly inspects treatment facilities and enforces permit requirements.

Enhancements to the regulation and management of municipal wastewater discharges to the Bay have occurred in phases. By the 1950s, many Bay Area communities had built primary sewage treatment plants, which removed material that could be screened or would either float or readily settle out by gravity. This low level of treatment left large amounts of pollutants flowing into the Bay, and the Bay suffered from high concentrations of fecal bacteria, low dissolved oxygen, frequent fish kills, foul odors, and other problems.

Along with the NPDES Program, the Clean Water Act provided clear goals and billions of dollars toward construction of Bay Area municipal wastewater treatment facilities. A few facilities, including those of San José-Santa Clara, Oro Loma, and Dublin-San Ramon, were providing secondary treatment by the late 1960s. Secondary treatment generally removes 80% to 90% of oxygen-demanding organic waste and suspended solids. Between 1960 and 1985, over $3 billion was spent in the Bay Area to upgrade and construct wastewater treatment plants, to move outfalls into deeper water, and to seasonally limit discharges to shallow water. By 1987, all municipal wastewater treatment plants discharging to the Bay were providing at least secondary treatment.

The widespread adoption of secondary treatment for Bay Area municipal wastewater drove a quantum leap of improved Bay water quality. By 1985, Bay Area municipal wastewater treatment plants had reduced suspended solids loading by 80% and biochemical oxygen demand (BOD) by 88% from the high levels recorded two decades earlier, while the service area population increased by 52% over the same period. Dissolved oxygen (DO) levels in the Bay increased in response to the reduced inputs of BOD. For example, long-term DO monitoring in the Lower South Bay spanning more than 50 years has shown an elimination of the low levels that were common from the late 1950s through the 1970s (**Figure 3**). In the late 1970s, the three Lower South Bay treatment plants contributed to the reductions in this region by installing nitrification and filtration to reduce oxygen demand.

Pretreatment Programs Reduce Pollution from Industrial Sources

After secondary treatment was established across the region, load reductions and Bay water quality improvement continued, driven by pretreatment and pollution prevention. Pretreatment and pollution prevention reduce pollution at its source, which is often more efficient than treating polluted wastewater at a treatment plant.

Although some large industrial facilities have their own wastewater treatment plants and discharge their treated effluent directly to the Bay or its tributaries under their own NPDES permits (**pages 22-35**), many other smaller industrial facilities discharge their wastewater indirectly, through municipal sewer systems. This industrial wastewater may contain a variety of harmful substances (such as industrial process by-products, like copper, lead, nickel and other heavy metals).

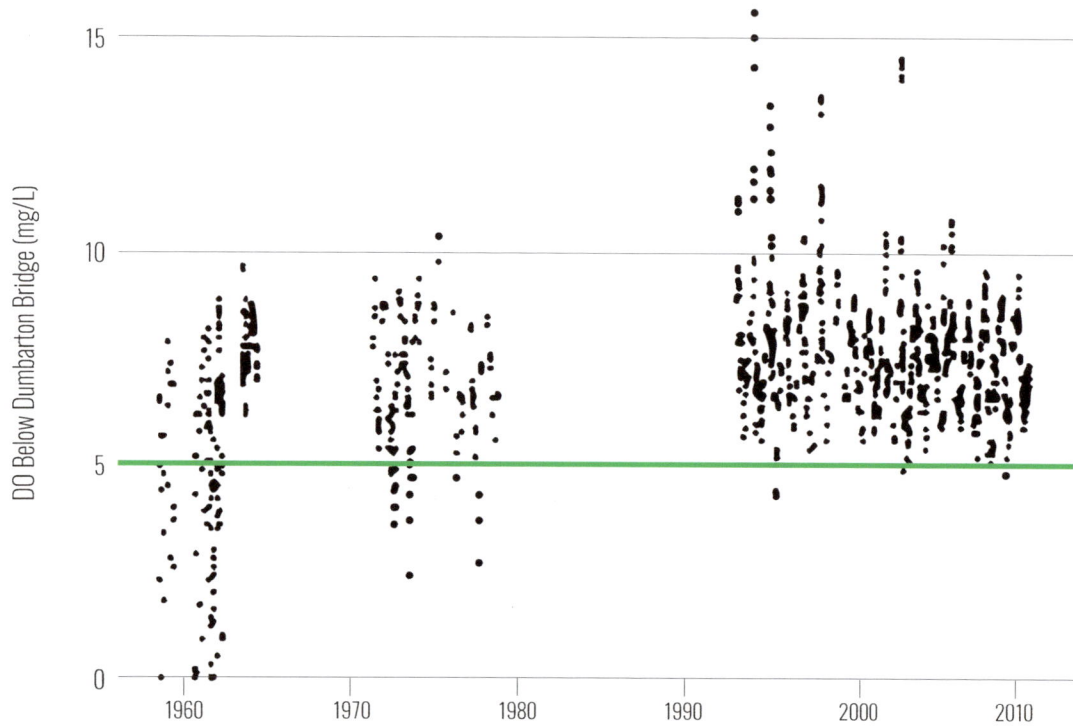

Figure 3. The widespread adoption of secondary treatment for Bay Area municipal wastewater after passage of the Clean Water Act in 1972 drove a quantum leap of improved Bay water quality. For example, dissolved oxygen (DO) levels in the Bay increased in response to the reduced inputs of organic waste. Long-term DO monitoring in the Lower South Bay spanning more than 50 years has shown an elimination of the low levels that were common from the late 1950s through the 1970s

FOOTNOTE: Dissolved oxygen south of the Dumbarton Bridge. The green line represents a common standard to protect marine fish sensitive to low oxygen. From Cloern and Jassby (2012).

Because sewage collection and treatment systems are not designed to remove all these substances, industrial waste can damage sewers and interfere with treatment plant operation, pass through the systems untreated, polluting nearby waters, and increase the costs and environmental risks of biosolids management. The practice of removing pollutants from industrial wastewater before it is discharged into municipal sewage treatment systems is known as "pretreatment."

The Clean Water Act established the framework for a national pretreatment program as a component of the NPDES program. In 1978, USEPA established regulations requiring many municipal wastewater agencies to develop and implement local pretreatment programs. USEPA delegated the responsibility to oversee these local pretreatment programs to the State Water Resources Control Board and Regional Water Boards in 1989, thus the Water Board approves the pretreatment programs of local municipalities, including permitting, administrative, and enforcement tasks to reduce industrial discharges into municipal wastewater treatment plants. In the San Francisco Bay Region, there are about 25 local pretreatment programs. The NPDES permits for these municipal wastewater treatment plants spell out the pretreatment program requirements.

Pollution Prevention Programs Reduce Pollution from Residential and Commercial Sources

While pretreatment focuses on industrial sources, pollution prevention focuses on residential and commercial sources. Municipal wastewater agencies implement pollution prevention programs to encourage residents and businesses to reduce wastewater pollution. Pollution prevention reduces or eliminates waste at the source by modifying production processes, promoting use of non-toxic or less-toxic substances, implementing conservation techniques, and re-using materials rather than putting them into a waste stream.

Pollution prevention is especially important for pollutants such as metals, plastics, and organic constituents that are not destroyed during wastewater treatment. Treatment at a wastewater facility is not a realistic solution for these constituents. Even if a wastewater treatment plant successfully removes wastewater pollutants from its effluent discharge, the pollutants are often simply transferred to biosolids. California's Department of Resources Recycling and Recovery (CalRecycle) is in the process of finalizing regulations (SB 1383) that will prevent most biosolids from being sent to landfills; therefore, most biosolids in California will need to be beneficially reused through some form of land application within the next few years.

The California Water Code authorizes the Water Board to require certain dischargers to develop pollution prevention plans. The Water Board first did this in 1988, and later expanded the program. In 1990, the Water Board collaborated with wastewater dischargers to form the Bay Area Pollution Prevention Group (BAPPG). The BAPPG now operates as an arm of the Bay Area Clean Water Agencies and includes 43 Bay Area wastewater agencies that meet bi-monthly to coordinate pollution prevention activities and leverage resources for smaller agencies. Broad regional participation makes region-wide projects possible. NPDES permits require municipal dischargers to continue their pollution prevention efforts.

Regional pollution prevention targets have included copper; fats, oils, and grease; mercury; silver; pesticides; pharmaceuticals; triclosan; trash; and wipes. Some examples of pollution prevention projects include outreach to dental professionals regarding best management practices for dental amalgam to reduce mercury discharges; coordination with the Department of Pesticide Regulation to obtain a prohibition on use of copper-based root control products and tributyltin cooling water additives in the Bay Area; Our Water – Our World, a program that raises awareness of the connection between pesticide use and water quality and provides information to consumers at the point-of-purchase about less-toxic alternatives; and outreach regarding safe pharmaceutical disposal, including pharmaceutical take-back programs.

Together, pretreatment and pollution prevention programs further reduced pollutant loadings after secondary treatment plant construction was completed. For example, copper and nickel loads from four large treatment plants (San José-Santa Clara, San Francisco, EBMUD, and Central Contra Costa Sanitary District) that discharge approximately half of the total volume of treated wastewater flowing to the Bay decreased by an additional 75% from 1986 to 2005, largely due to pretreatment and pollution prevention programs.

Monitoring Optimized to Meet Regional Needs

Treatment plant discharge and receiving water monitoring are critical components of the NPDES program, allowing assessment of the effects of discharges on receiving waters and ensuring compliance with NPDES permit requirements. Since 1993, the Water Board has used NPDES permits and other regulatory instruments to require Bay Area dischargers to support the Regional Monitoring Program for Water Quality in San Francisco Bay (RMP). Now in its 27th year, the RMP informs stewardship of the Bay with one of the best water quality monitoring programs in the world (see Davis 2017 for a historic overview of the Program). Municipal wastewater agencies, in cooperation with other discharger groups, support the RMP with funding and actively participate in Program governance. In 2016, the Water Board significantly modified NPDES permit monitoring requirements to establish "alternate monitoring requirements" that allow dischargers the option to redirect funds from less useful monitoring they had conducted individually to more useful monitoring conducted by the RMP. This enhanced funding (approximately $280,000 per year) allows the RMP to pursue more proactive and adaptive monitoring of contaminants of emerging concern (CECs), with a goal of early detection and prevention of potential problems before any significant impacts on Bay water quality occur.

Watershed Permits Provide Consistency Across the Region

The Water Board has also issued two region-wide, or "watershed," NPDES permits that apply to multiple discharges. These innovative permits provide for a coordinated approach for regulating the many treatment plants within the San Francisco Bay watershed.

In 2007, the Water Board adopted a watershed permit that addresses discharges of mercury to the Bay from more than 40 municipal dischargers and 10 industrial dischargers. PCBs were added to the watershed permit in 2012. This permit implements loading limitations established in Total Maximum Daily Load (TMDL) control plans developed by the Water Board. The Water Board reissued this permit in 2017.

In 2014, the Water Board issued a watershed permit for nutrient discharges from the more than 40 municipal wastewater treatment plants in the region. The Water Board reissued this permit in 2019. This permit seeks to prevent potential future water quality harm if the Bay's current resilience to high nutrient loads diminishes. The reissued permit requires increased municipal discharger support for scientific studies to characterize San Francisco Bay's response to nutrient loads to $2.2 million per year. Further, it requires municipal dischargers to evaluate opportunities to reduce nutrient discharges using "green" solutions, like natural systems (e.g., wetlands) and wastewater recycling – opportunities that can provide multiple benefits beyond nutrient removal (e.g., resilient water supplies, protection against sea-level rise, and removal of contaminants of emerging concern).

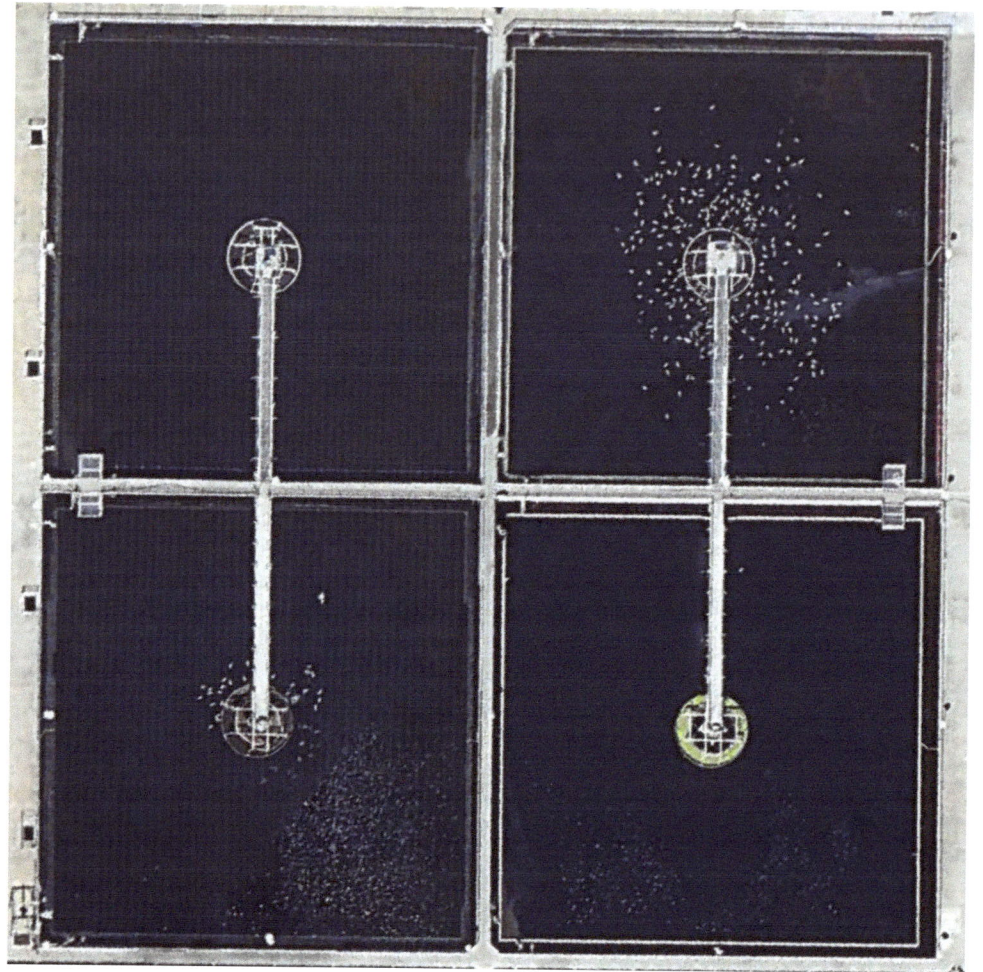

▲ **Clarifying ponds at the Palo Alto Regional Water Quality Control Plant** (Google Earth)

Recent Trends and Findings

Human activities and human waste are the primary sources of the pollutants that flow into Bay Area municipal treatment plants. The number of humans in the Bay Area has been steadily increasing ever since the Gold Rush era (**Figure 4**). From 1970 to 2010, the population of the Bay Area counties increased by over 50%, from 4.6 million to 7.2 million. By 2018, according to US Census Bureau data, the population grew to 7.8 million, and the Bay Area had the fastest population growth rate of any region in California (Green and Shuler 2019). The growth from 2010-2018 (0.6 million people) was higher than that observed from 2000-2010 (0.4 million people), and thought to be driven by the strong Bay Area economy. It is estimated that by 2040 the population will increase by another 1.8 million people to reach 9.6 million (MTC and ABAG 2017).

In spite of this long-term trend of population increase, the flow of treated effluent to the Bay from municipal wastewater treatment plants has declined since 1997 (**Figure 5**). Rainfall patterns have influenced the flow trend because periods of low rainfall lead to less stormwater and groundwater flow into the sewage conveyance system through leaky pipes and manholes. In addition, periods of drought lead to increased water conservation. These factors contributed to lower municipal wastewater flows during the five-year drought from 2012-2016, followed by relatively high flows in 2017 when a wet winter ended the drought. Nevertheless, the potential for flows to increase in the future will be limited to the extent that available water supplies may not keep up with population growth.

Similar to flows, loads of BOD and metals have remained low since the 1980s due to secondary treatment, pretreatment, and pollution prevention (as discussed in the previous section), and the impacts of these pollutants on Bay water quality remain under control.

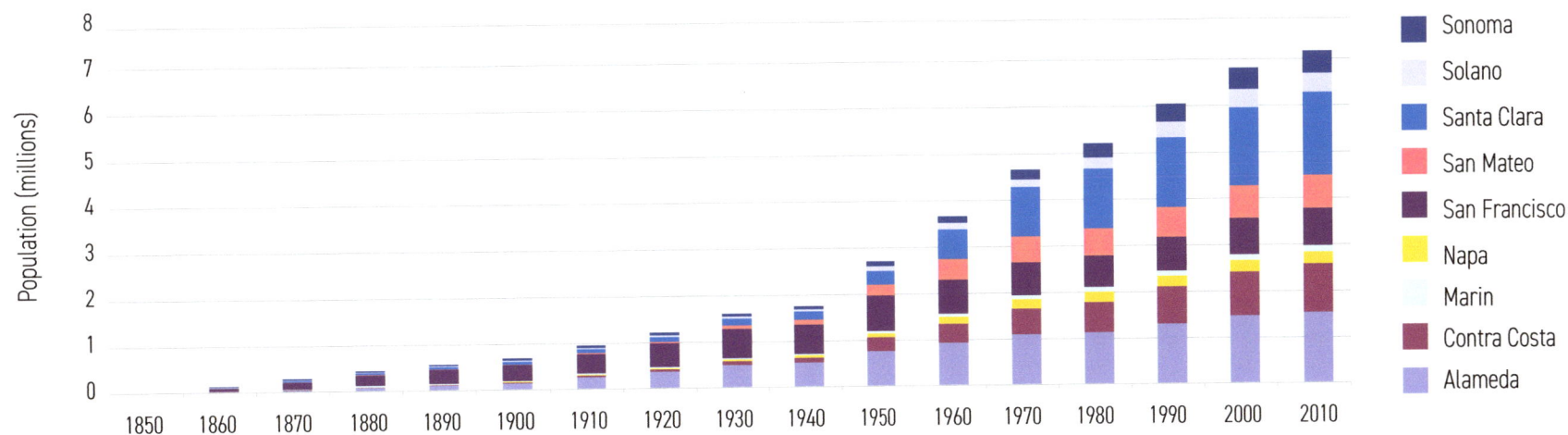

Figure 4. From 1970 to 2010, the population of the Bay Area counties increased by over 50%, from 4.6 million to 7.2 million. By 2018, according to US Census Bureau data, the population grew to 7.8 million, and the Bay Area had the fastest population growth rate of any region in California, likely driven by the strong Bay Area economy. By 2040 the population will increase by an estimated 1.8 million people to reach a total of 9.6 million.

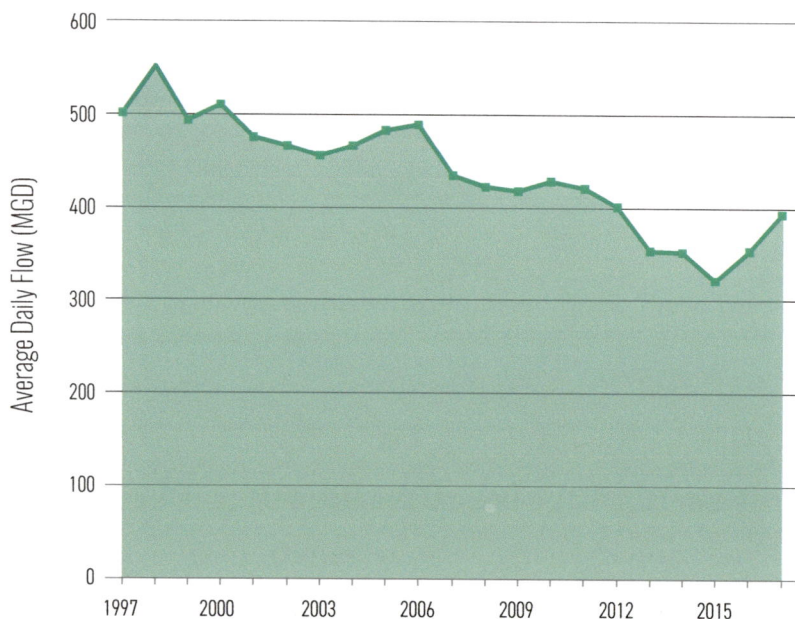

Figure 5. In spite of the long-term trend of population increase, the flow of treated effluent to the Bay from municipal wastewater treatment plants has declined since 1997. Rainfall patterns have influenced the flow trend because periods of higher rainfall lead to more stormwater and groundwater flow into the sewage conveyance system through leaky pipes or manholes. The larger average flow in 2017 was driven by the wet winter that ended a five-year drought from 2012-2016.

FOOTNOTE: Based on annual average (calendar year) flows to the Bay from the ten largest municipal wastewater treatment plants: San Jose, East Bay Dischargers, East Bay MUD, San Francisco, Central Contra Costa, Palo Alto, Fairfield-Suisun, Silicon Valley Clean Water, San Mateo, and Sunnyvale.

Concentrations of mercury and PCBs in some Bay fish species are above targets established to protect the health of humans and wildlife. Because of these high levels, a fish consumption advisory is in effect for the Bay, and TMDL control plans have been established. The TMDLs for mercury and PCBs have determined the overall load of these pollutants that the Bay can receive and still meet the regulatory targets, and each major pollutant pathway has been allocated a portion of this overall load ("wasteload allocations"). Substantial reductions in overall load are needed for both mercury and PCBs. However, municipal wastewater contributes a small proportion of the total load for these contaminants (0.6% for mercury and 2.6% for PCBs), and the current loads are well below the municipal wastewater wasteload allocations (**Figure 6**). A TMDL has also been established for selenium in the North Bay, driven by concern for the health of white sturgeon. However, similar to mercury and PCBs, municipal wastewater contributes a small proportion (2%) of the total selenium load to the North Bay.

Other pollutants, however, threaten to have significant negative impacts on Bay water quality. As mentioned in the previous section in regard to the Nutrient Watershed Permit for municipal wastewater discharges, ensuring that nutrient loads to the Bay do not begin to cause problems related to algal blooms is a high priority for water quality managers. Municipal wastewater is a dominant contributor of nutrients to the Bay. For nitrogen, the nutrient of highest concern, recent data indicate that Bay Area municipal treatment plants are contributing 62% of the load to the Bay as a whole (**Figure 6**). Data show that a significant upgrade in San José's wastewater treatment process in the late 1990s drove a 40% decrease in nitrogen concentrations in the Lower South Bay (**Figure 7**). More recent data, however, indicate that nitrogen loads to the Bay have increased over the last six years (**Figure 8**). Data from 2012 through 2018 indicate that the overall total nitrogen load from all Bay Area municipal wastewater discharges has trended upward, from an average of 53,000 kg N/d in 2012/2013 to 57,000 kg N/d in 2017/2018 (BACWA 2018). The suspected drivers of this 5% increase are the growing population, economic conditions that attract workers to the Bay Area, and the increasing importation of organics to wastewater treatment plants for energy production (discussed further below). Given the projected continued increase in Bay Area population, the recent trend data suggest that nitrogen loads are likely to rise by 1-2% per year unless nutrient removal is further enhanced.

Mercury

- Atmospheric Deposition (6%)
- Guadalupe River (20%)
- Other Nonurban Stormwater (6%)
- Municipal Wastewater (0.6%)
- Industry (0.1%)
- Urban Stormwater (26%)
- Delta (42%)

Nitrogen

- Delta (22%)
- Municipal Wastewater (62%)
- Refinery (1%)
- Urban Stormwater (15%)

PCBs

- Stormwater (69%)
- Industry (0.4%)
- Municipal Wastewater (3%)
- Delta (29%)

Selenium

- Delta (77%)
- Municipal Wastewater (2%)
- Refinery (11%)
- Urban and Nonurban Stormwater (10%)

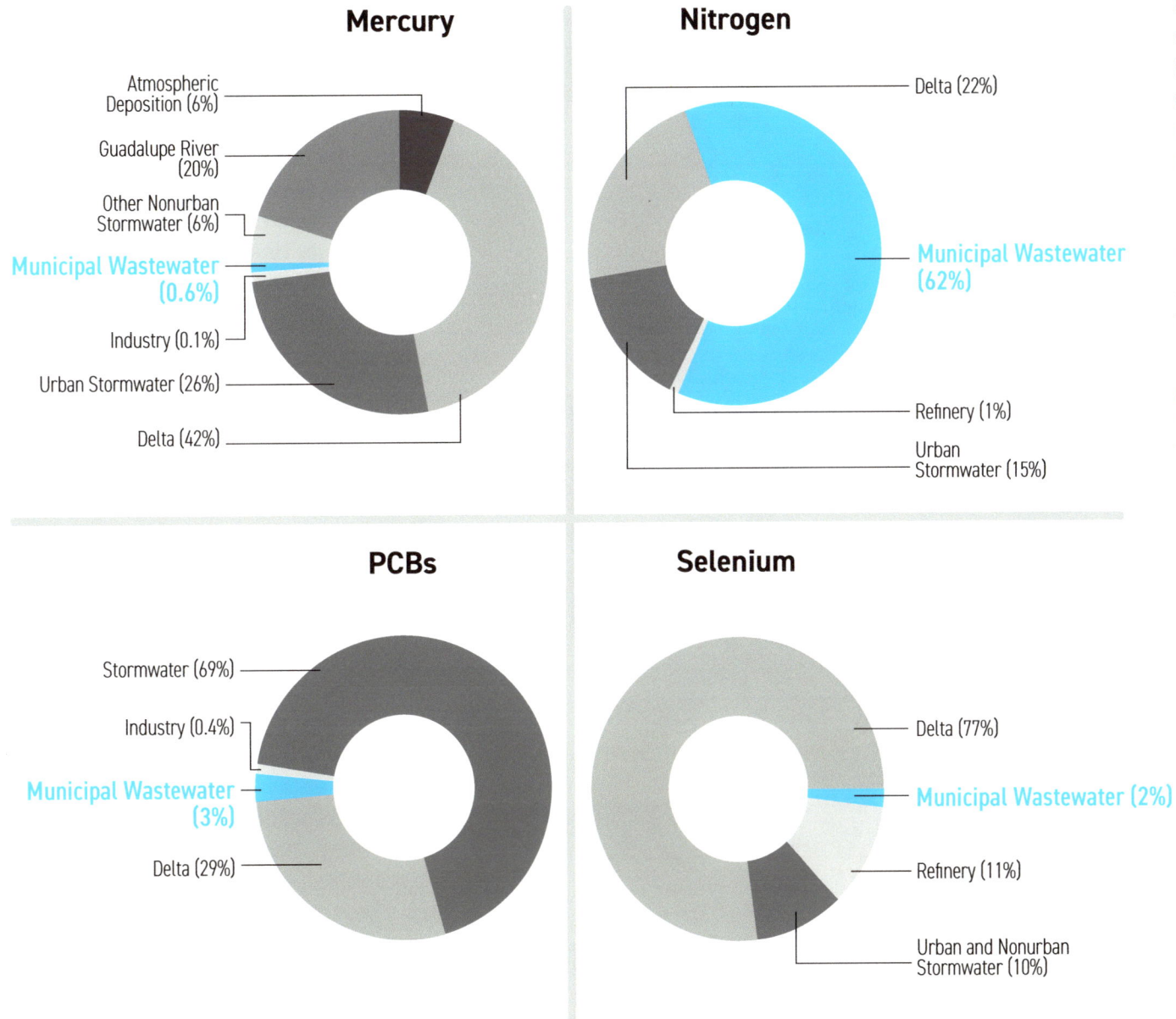

Figure 6. Mercury, PCBs, selenium, and nitrogen have been a focus of regulatory attention and the subject of inventories of loading to the Bay. Municipal wastewater is a dominant pathway for nitrogen (62% of the overall load), but contributes less than 3% of the loading of the other three pollutants.

FOOTNOTE: Pathway categories vary by pollutant because they were treated differently in the TMDLs.

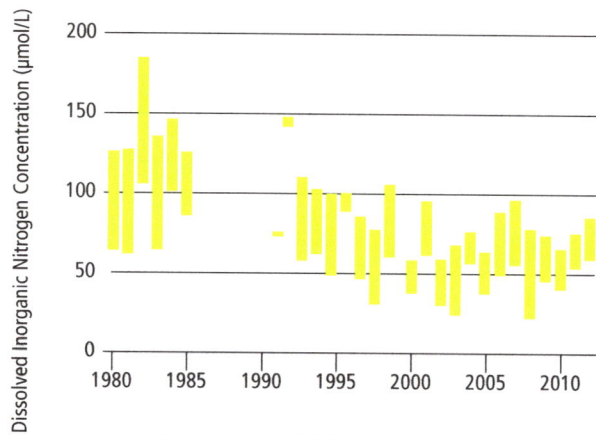

FOOTNOTE: After Novick et al. (2017). DIN concentration ranges are based on year-round data from surface samples at multiple stations in Lower South Bay (USGS s34-s36) and South Bay Dischargers Authority (SBDA) monitoring station SB5, also located in the main channel of LSB. The box extends from the 25th percentile to the 75th percentile. Data are from the USGS water quality component of the RMP and available online (Schraga and Cloern 2017).

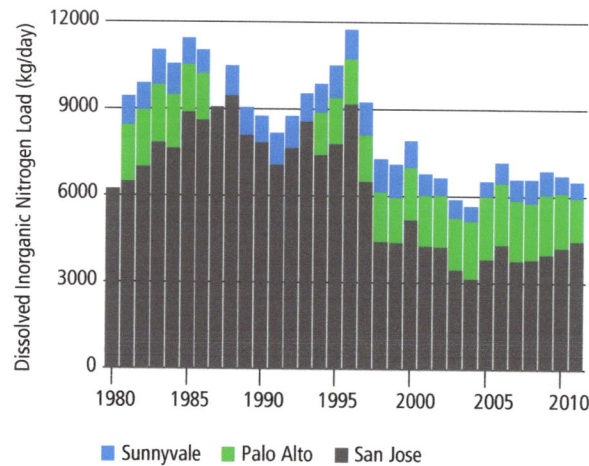

FOOTNOTE: Loads were calculated based on reported effluent flow/concentration data by the three major dischargers to LSB. There are some years where data was not available for Sunnyvale and Palo Alto.

Figure 7. A significant upgrade in the wastewater treatment process for the largest municipal discharge in the Bay – the San José-Santa Clara Regional Wastewater Facility - in the mid-1990s sharply reduced the overall nitrogen loading to the Lower South Bay. This load reduction drove a 40% decrease in nitrogen concentrations in the Lower South Bay, demonstrating the potential effectiveness of load reductions.

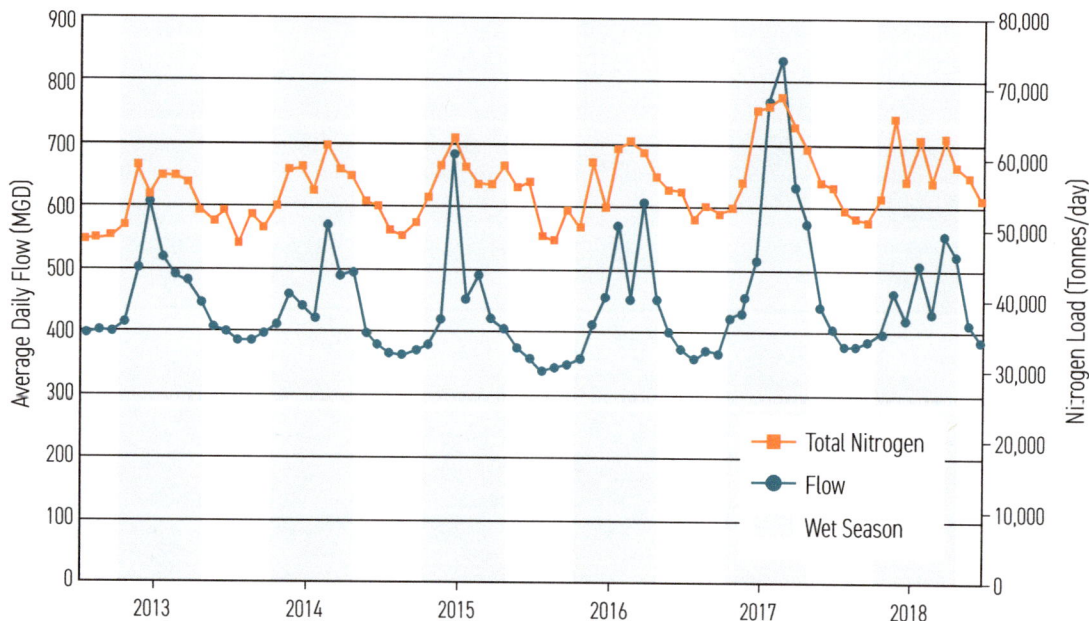

Figure 8. More recent data, however, indicate that nitrogen loads to the Bay have increased over the last six years. Data from 2012 through 2018 indicate that the total nitrogen load from Bay Area municipal wastewater discharges has increased by 5%. The suspected drivers are the growing population, economic conditions that attract workers to the Bay Area, and the increasing importation of organics to wastewater treatment plants for energy production. Given the projected continued increase in Bay Area population, the recent trend data suggest that loads are likely to rise by 1-2% per year unless nutrient removal is further enhanced. From BACWA (2018).

Municipal wastewater is also an important pathway for many CECs. The RMP uses a tiered framework to prioritize the multitudes of potential CECs (**page 76**). The highest priority CECs fall into the "moderate concern" tier, where concentrations observed in the Bay suggest a high probability of a low-level effect on Bay aquatic life. Municipal wastewater is a major pathway for all of the contaminants in the moderate concern tier, including perfluorooctane sulfonate (PFOS) and certain other related fluorinated chemicals (primarily used as stain-repellents); the pesticides fipronil and imidacloprid; alkylphenols and alkylphenol ethoxylates (common ingredients of detergents and cleaning products); bisphenols (plastic additives); organophosphate esters (flame retardants and plastic additives), and microplastics.

A recent monitoring study (Sadaria et al. 2016) and literature review (Sutton et al. 2019a) showed that pesticides reach the Bay via municipal wastewater. Part of the problem with pesticides in wastewater is that conventional wastewater treatment is generally ineffective at removing many of them, with high removal efficiency only observed in the case of highly hydrophobic compounds, such as pyrethroids, that attach to particles and are removed via solids removal processes.

Municipal wastewater is also a pathway for microplastics to reach the Bay. Microplastic loading to the Bay and levels in Bay water, sediment, and biota have been examined in a groundbreaking study funded primarily by the Moore Foundation, with additional support from the RMP and others (Sutton et al. 2019b). Based on a subset of particles that were subject to spectroscopy to identify their chemical composition and discharges from eight facilities representing 70% of the total discharge to the Bay, the study estimated approximately 17 billion microplastic particles are discharged annually to the Bay from wastewater treatment facilities. This is substantially lower than the estimate developed for small tributaries surrounding the Bay (7 trillion microplastic particles per year). The complete report will be released this fall (Sutton et al. 2019b).

Municipal wastewater is the primary pathway to the Bay for pharmaceuticals, which are as a class categorized as a "low concern" in the RMP tiered framework. Monitoring of municipal wastewater effluent in 2016 and 2017, however, identified 17 compounds as priorities for further evaluation, including six antibiotics, three antidepressants, an anti-convulsant, three painkillers, an antihistamine, an antidiabetic, and two medications for high blood pressure.

Future Work, Directions, and Challenges

Because many of the wastewater facilities in the Bay Area and nationwide date back to the 1972 Clean Water Act and subsequent federal construction grants, they are reaching the end of their useful lifespans. Due to their age, as well as the need to meet updated seismic standards and protect facilities against sea level rise, a good percentage of wastewater agencies in the region are looking at investing in major facility rebuilds or rehabilitations over the next several years. Agencies such as the Cities of Sunnyvale and San Mateo, both currently involved in major construction projects, are also using this opportunity to implement new treatment technologies that will enhance nutrient removal. The San José-Santa Clara Regional Wastewater Facility is in the middle of a 10-year, $1.4 billion capital improvement program that will rehabilitate and improve the efficiency, reliability, and sustainability of all stages of its treatment processes.

At the same time as wastewater facilities are recommitting to continuing and exceeding the high level of treatment they have been achieving over the past several decades, new societal and environmental pressures are leading wastewater agencies to re-envision themselves as resource recovery agencies. Wastewater agencies are providing increased volumes of recycled water to offset potable water consumption, generating carbon-neutral energy, and exploring ways to enhance sea level rise protection and provide habitat by creating wetlands. Low flows resulting from drought and conservation are driving changes in the assumptions that have been used to design and operate facilities. Efforts to address these challenges and opportunities often come with unintended consequences or cross-media impacts that can be challenging to address.

Potable Reuse is Increasingly the Future of Water Recycling

Bay Area water and wastewater agencies have long been working to increase the recycled water produced for landscape and agricultural irrigation, as well as industrial reuse. However, urban "purple pipe" projects, where recycled water is used for non-potable uses, are very expensive due to the need to build a new distribution system, separate from the potable system. Additionally, since many of the lower-cost non-potable recycled water projects in the denser inner urban areas have already been constructed, a consensus is being reached among water managers that the future of water recycling is potable reuse.

The State Water Resources Control Board is developing criteria for potable reuse that may take until 2025 to be finalized. However, the current practice is that any treatment train for potable reuse should incorporate reverse osmosis to protect public health. Reverse osmosis works by pushing source water, in this case treated wastewater effluent, at high pressure through a membrane with extremely fine pores. Most of the pollutants, and about 15% of the water, make up the "reverse osmosis concentrate," which is the byproduct of the process. This reverse osmosis concentrate, which contains roughly the same pollutant load as the source water but at higher concentrations due to the reduced water volume, poses a management concern.

NPDES permits regulate most pollutants based on concentration, rather than by load, because the toxicity of many pollutants depends on their concentrations in the receiving water. Accepting reverse osmosis concentrate to discharge through their outfalls may be a compliance risk for wastewater agencies, particularly those that discharge to shallow waters and therefore get little or no dilution credit. Projects producing reverse osmosis concentrate, unlike recycled water irrigation projects with less advanced treatment, will not reduce loading of nutrients or other pollutants to the Bay. The Silicon Valley Advanced Water Purification Center, mentioned previously, is working with researchers at Stanford and UC Berkeley to investigate different alternatives for the removal of metals, nutrients, and emerging contaminants from reverse osmosis concentrate.

Waste-to-Energy Programs Mitigate Climate Change, But Increase Nutrient Discharges

An exciting development over the past decade is the development of waste-to-energy programs at wastewater treatment plants. Most wastewater facilities use anaerobic digesters to produce biosolids, a stabilized product of the solids generated in the wastewater treatment process. The biosolids are often used at landfills for alternative daily cover (a practice that will likely be discontinued in the next few years to comply with the new regulations in SB 1383) or applied to agricultural land. As industries have left the Bay Area, reducing organic loads to treatment plants, some wastewater agencies have been left with excess digester capacity. In addition to digesting their own solids derived from wastewater treatment, several facilities now receive food or agricultural waste that may have otherwise been landfilled. The digestion of organic waste produces methane, which can be burned to make electrical energy. Because the source of this fuel is biogenic organic waste, rather

than anthropogenic fossil fuels, these waste-to-energy programs are an important component by which the state can meet its climate change mitigation goals.

While these co-digestion projects produce a societal benefit by producing non-fossil fuel derived energy, the nutrient load in the food or agricultural waste is added to the load otherwise discharged by the wastewater facility. Treating the high-nutrient sidestream from digested solids dewatering can be cost effective compared to other nutrient treatment strategies.

Wastewater Can Provide Fresh Water to Enhance Wetlands for Sea Level Rise Protection

As sea levels rise and storm intensities increase, wetlands are being considered as a multi-benefit solution to protect near-shore areas. Wetlands may either be part of the receiving water, and regulated accordingly, or part of the treatment train within the wastewater treatment plant. Since wastewater treatment plants are often located in low-lying areas, protecting these facilities with wetlands sustained by wastewater effluent is a compelling management strategy.

Wetlands sustained by treated wastewater provide many benefits.

- Freshwater/brackish wetlands support more diverse habitat than saltwater wetlands.

- Treatment wetlands can assimilate nutrients, metals, and CECs, with lower energy and chemical demands than conventional treatment.

- Wetlands can mitigate the impacts of sea level rise by increasing shoreline resiliency.

- Wetlands may improve the ability of native wildlife species to withstand longer and more severe droughts within a changing climate.

Oro Loma Sanitary District has pioneered the concept of the horizontal levee, where subsurface flow of treated wastewater sustains wetlands with tall grasses that enhance the wave dampening function of the levee and reduce erosion. Concurrently, the treated wastewater is further polished through the wetland, and high levels of nutrient and CEC removal are observed. Oro Loma is looking to expand its horizontal levee, and other agencies are considering similar projects in the North Bay and South Bay. As part of the Nutrient Watershed

Permit, wastewater agencies will be looking around the Bay for opportunities to implement wetlands projects that also reduce nutrient loads to the Bay.

Conservation Has Unintended Consequences

Bay Area residents were very successful at implementing measures to conserve water during the 2012-2016 drought. Since much of the conservation involved installation of low-flow fixtures, water use levels are not expected to return to predrought levels. Conservation has resulted in lower flows with "higher strength" (higher pollutant concentrations). The lower flow in collection systems that convey sewage to wastewater facilities can lead to blockages in sewer pipes if the flows do not achieve flushing velocities. In addition, in collection systems receiving lower flows than they were designed for, high-strength flows often become anaerobic, leading to both corrosion and odor problems. During the drought, wastewater agencies throughout the state also observed higher levels of corrosion in their headworks. The biological and chemical reactions in collection systems also affect treatment processes, as well as the ability to digest the solids and collect methane for co-digestion and energy production. Another impact of lower flows is that less water is available for recycled water.

Into the Future

The coming decades will see shifts in how municipal wastewater treatment plants are envisioned in the community. They will no longer be viewed simply as facilities for cleaning up wastewater prior to discharge, since wastewater agencies around the Bay Area are now considering multi-benefit projects using wastewater to produce recycled water, renewable energy, and soil amendments to sequester carbon, as well as using treated effluent to enhance habitats in the Bay margins to protect against sea level rise. These new projects will need to be balanced with the mission of protecting water quality in San Francisco Bay and public health in the face of a growing population, and controlling sewage collection and treatment rate increases in a region that already has one of the highest costs of living in the world. To this end, wastewater agencies will look to secure new funding sources at the local, state, and federal levels to finance projects that will enhance the environment, protect infrastructure in a changing climate, and deliver benefits to the communities they serve. §

Wastewater treatment plants will no longer be viewed simply as facilities for cleaning up wastewater prior to discharge

▲ **Reverse osmosis array at the Silicon Valley Advanced Water Purification Center** (Valley Water)

INDUSTRIAL WASTEWATER

Jay Davis
San Francisco Estuary Institute

John Madigan and **Robert Schlipf**
*San Francisco Bay
Regional Water Quality Control Board*

Bridgette DeShields
Integral Consulting

Peter Carroll
Marathon Petroleum Corporation

Maureen Dunn
Chevron

Kevin Buchan
Western States Petroleum Association

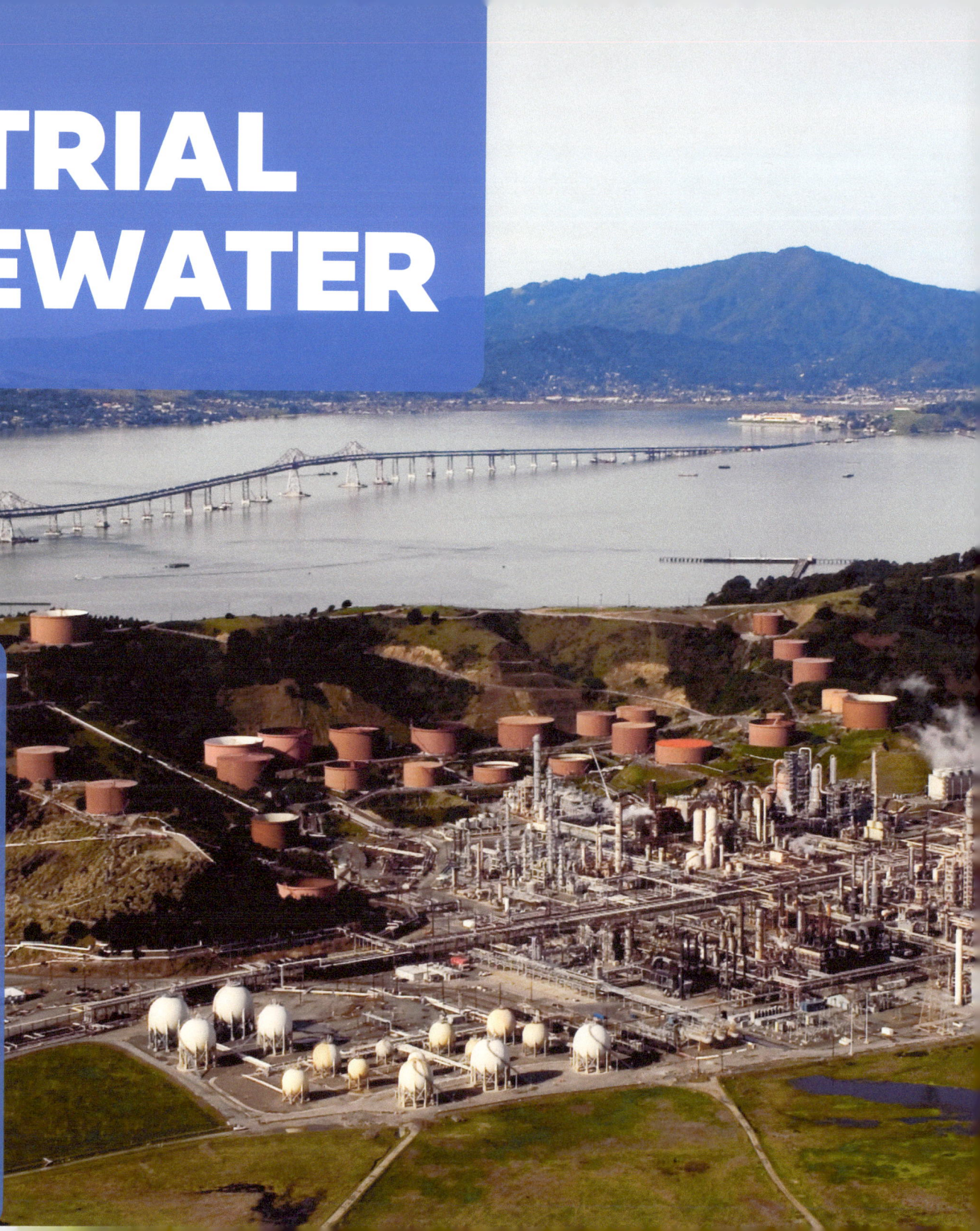

HIGHLIGHTS

▶ Contaminants in industrial wastewater waste streams are largely removed by wastewater treatment plants

▶ Industrial wastewater is a minor source at the regional scale for most pollutants of concern in the Bay

▶ Industrial wastewater is highly regulated in the San Francisco Bay Area, with increasingly stringent regulations driving the adoption of secondary treatment for petroleum refineries in the 1970s, and activated carbon treatment and selenium removal in the 1990s

▶ Five petroleum refineries account for most of the flow and pollutant loading from direct industrial discharges to the Bay

▶ Although industrial wastewater is a source of selenium to San Francisco Bay, the largest selenium load to the Bay is from the Delta

▶ The Bay Area refineries have played a central role in improving understanding of North Bay selenium in support of TMDL development and implementation

▶ Industrial wastewater infrastructure planning needs to address changing regulations, sustainable water use, and climate change

◀ **The Chevron refinery at Point Richmond** (Alamy)

Industrial Wastewater 101

Bay Area industries provide many goods that we depend on in our daily lives, including food, beverages, clothes, energy, building materials, and paper and chemical products, and water is a key component of the production processes for many of them. Petroleum refining is a particularly important industrial activity in the Bay Area. Five refineries are located in the North Bay, in a 17-mile stretch between Richmond and Benicia (**Figure 1**). Other prominent industrial facilities in the region include C&H Sugar in Crockett and USS-POSCO (a steel finishing plant) in Pittsburg.

The petroleum refineries are a large contributor to the Bay Area economy. California ranks third in the nation in petroleum refining capacity. The state's 17 refineries – located in the Central Valley, Los Angeles County, and the Bay Area – have a combined capacity of nearly two million barrels per day. Bay Area refineries account for a large proportion of the statewide total petroleum product market demand, with an average of 760,000 barrels per day. The petroleum is refined to create a wide variety of products, including gasoline, diesel fuel, jet fuel, fuel oil, and asphalt, as well as chemicals that are ingredients in plastics, paint, roof shingles, cosmetics, candles, shampoo, and tires. The five refineries sustain 88,000 jobs in the Bay Area and have an estimated annual economic impact of $78 billion.

Industrial production processes and other activities in industrial facilities, such as cleaning and cooling, generate wastewater that may contain pollutants. Most industrial facilities discharge their wastewater to municipal sewage collection systems, where it flows to a municipal wastewater treatment plant. The Pretreatment Program (**page 11**) requires these "indirect" industrial dischargers to minimize the release of pollutants that could interfere with municipal wastewater treatment or adversely affect Bay water quality. A relatively small number of industrial facilities, including the refineries, C&H Sugar, USS-POSCO, and

a few others, have their own wastewater treatment plants and discharge permits, and discharge directly to the Bay or its tributaries. This treated wastewater is strictly regulated so it can be safely discharged into water bodies, applied to land, or reused in plant operations. This article focuses on these direct industrial discharges, and, specifically, discharges from the petroleum refineries.

In contrast to municipal wastewater treatment facilities, which are pollutant pathways merely conveying pollutants from households and businesses to the Bay after substantial treatment, industrial facilities and their wastewater treatment plants represent sources: activities that introduce pollutants into waste streams that are then discharged, also after substantial treatment, into the environment. These industrial facilities directly control the processes that lead to the presence of pollutants in wastewater streams and can implement source reduction and pollution prevention measures to minimize loadings to the Bay.

In the case of the petroleum refineries, crude oil processing is the source of most of the pollutants that enter the wastewater stream (**page 26**). Crude oil is a fossil fuel, formed when large quantities of algae, zooplankton, plants, and other living organisms were buried by sediments that formed sedimentary rock and were subjected to intense heat and pressure over millions of years. The elemental composition of crude oil mirrors that of living organisms, chiefly composed of carbon and hydrogen, with smaller amounts of nitrogen, oxygen, sulfur, and trace elements. Crude oil is a complex mixture of hundreds of different hydrocarbons (including alkanes, cycloalkanes, and aromatics) and other chemicals, some of which can be toxic.

Like the municipal wastewater dischargers (**pages 2-21**), the direct industrial dischargers make substantial investments in infrastructure and labor to treat their wastewater to meet standards and protect Bay water quality. The refinery wastewater treatment plants have varying configurations, and employ the same basic elements as in treatment of

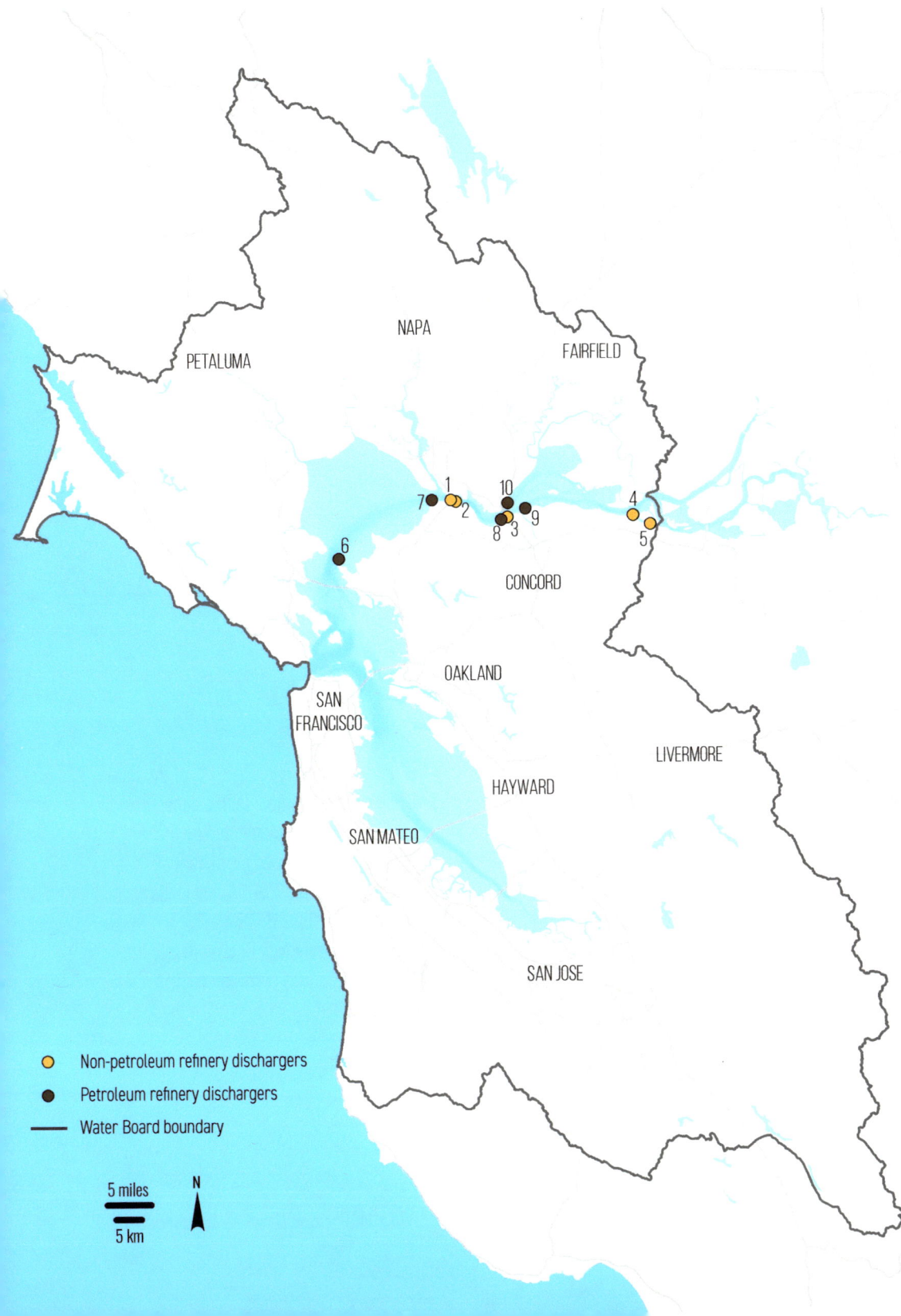

Non-petroleum refinery dischargers

1. C&H Sugar/Crockett Community Services District
2. Crockett Cogeneration LP/Pacific Crockett Energy
3. Eco Services Operations LLC
4. GenOn Delta, LLC
5. USS-Posco Industries

Petroleum refinery dischargers

6. Chevron Products Company
7. Phillips 66
8. Shell Oil Products US/Equilon Enterprises LLC
9. Marathon Petroleum
10. Valero Refining Company

Figure 1. Locations of 10 direct industrial discharges to San Francisco Bay. Five petroleum refineries are located in the North Bay, in a 17-mile stretch between Richmond and Benicia. Other prominent industrial facilities in the region include C&H Sugar in Crockett and USS-POSCO (a steel finishing plant) in Pittsburg.

⬤ Non-petroleum refinery dischargers
⬤ Petroleum refinery dischargers
— Water Board boundary

5 miles
5 km

N

Industrial Wastewater: Sources, Pathways, Loading

Industrial facilities are **SOURCES** of pollutants because their activities introduce pollutants into waste streams that are discharged, after substantial treatment, into the Bay. Petroleum refineries account for most of the industrial wastewater discharged to the Bay. Crude oil processing is the source of most of the pollutants that enter the refinery wastewater stream. Crude oil is a complex mixture of hundreds of different hydrocarbons, selenium, and other chemicals, some of which can be toxic.

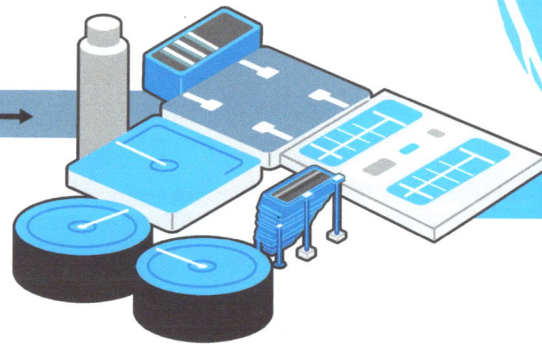

Refinery effluent contributes more than 1% of the total regional **LOADING** of only one pollutant: selenium. Selenium removal measures implemented by the refineries in the late 1990s yielded a substantial reduction in loading by the mid-2000s, and loads have remained at similar levels since that time. For the North Bay – the portion of San Francisco Bay subject to the Selenium TMDL – the refineries currently contribute 11% of the load.

Industrial dischargers make substantial investments to treat their wastewater to meet standards and protect Bay water quality. Refinery wastewater treatment plants employ the same basic elements as in treatment of municipal wastewater (settling and biological treatment), but also use other processes to remove oil, hydrocarbons, and selenium from the waste stream.

Illustration by Linda Wanczyk (lindawanczyk.com)

municipal wastewater (settling and biological treatment), but also use other processes that are specific to treatment of refinery wastewater. The plants generally include the following sequence of common elements.

- Sour water strippers use steam to remove ammonia and hydrogen sulfide.

- Oily water and solids separators remove oil and suspended solids by gravity. Chemicals may be added to coagulate and flocculate solids to expedite settling and removal processes. Dissolved air or dissolved nitrogen flotation may be included as another polishing step: flocculated solids and oil float to the surface and are mechanically removed.

- As with municipal plants, biological treatment is used to break down organic matter via microbial metabolism. Aeration supplies oxygen for the microbes and provides mixing. Nutrients may also be added to support microbial activity. Some plants add powdered activated carbon to the aeration cells to adsorb toxicants.

- Clarifiers then settle out biological solids, inert solids, and spent powdered activated carbon. Coagulants and flocculants may be added to enhance settling.

- Sand filters trap residual suspended particles and bacteria as a polishing step.

- Granular activated carbon filters (like Brita filters used in homes) or powdered activated carbon may be used to adsorb hydrocarbons and some metals, and are especially effective in removing contaminants that can be toxic to aquatic life.

- At refineries where sanitary waste is also treated, wastewater is disinfected using sodium hypochlorite, pH control, or other means to ensure removal of potentially pathogenic bacteria.

▲ **Oil tanker at Chevron Long Wharf, Point Richmond** (Alamy)

Similar to municipal wastewater treatment, refinery wastewater treatment has advanced in phases over the last several decades. Prior to the 1960s, simple treatment technologies such as gravity separators were employed to treat wastewater. In order to meet increasingly stringent environmental regulations, additional and more sophisticated control technologies were developed and implemented by the refineries. Around the mid-1960s and 1970s, the addition of aerated biological treatment ponds marked the beginning of significant upgrades that have continued to the present. In the 1970s, under the Clean Water Act, the USEPA proposed Best Practicable Technology guidelines for refineries. Several treatment upgrades were widely implemented in the years that followed. These upgrades included adding or expanding aeration ponds and tanks and installing clarification basins and deep water outfalls with diffusers. In the 1990s, all refineries in the San Francisco Bay Area installed activated carbon treatment steps (either with granular activated carbon or powdered activated carbon) to meet acute toxicity discharge permit limitations. In the most recent major treatment advance, steps to remove selenium were added by the refineries in the late 1990s.

Selenium in trace amounts is essential for cellular function in many organisms, including all animals, but high levels can be toxic. Fossil fuels like crude oil are derived from living organisms and therefore contain selenium. Selenium is a particular concern in the North Bay, where the refineries are located and where white sturgeon are abundant and at risk because they consume invasive clams that tend to accumulate high concentrations. Studies in the 1980s indicated that refineries were one of the primary sources of selenium to the North Bay. The refineries worked with the San Francisco Bay Regional Water Quality Control Board (Water Board) to reduce their selenium loads. In the late 1990s, several of the refineries installed Selenium Removal Plants (SRPs) to achieve reductions. SRPs use chemical precipitation and filtration to remove selenium. Precipitated selenium-containing solids are dewatered and sent to permitted waste management facilities. Other refineries used different methods to achieve load reductions, including a treatment wetland and more extensive treatment via oxidation ponds. The refineries were quite successful in achieving reductions (discussed further below).

The refineries and other industrial dischargers strive to ensure that the water that is discharged to the Bay meets the rigorous standards required by the Water Board. Extensive monitoring of effluent, upstream checks (for early warning), operator vigilance, and strong maintenance programs all work towards this end.

> Similar to municipal wastewater treatment, refinery wastewater treatment has advanced in phases over the last several decades

Regulatory Framework

As mentioned above, several San Francisco Bay Region industrial facilities discharge treated wastewater directly to surface waters under individual National Pollutant Discharge Elimination System (NPDES) Permits; multiple, similar smaller facilities discharge under general NPDES permits. The most significant industrial dischargers are the five petroleum refineries discharging to the Bay. The NPDES permits issued by the Water Board to industrial dischargers implement the federal Clean Water Act (CWA) requirement that wastewater discharges meet technology-based treatment requirements at a minimum, and any more stringent effluent limits necessary to meet water quality standards. Also pursuant to the CWA, the Water Board develops Total Maximum Daily Load (TMDL) control plans for waters that do not meet one or more water quality standards and implements them through effluent limits in new or existing NPDES permits.

Technology-based Effluent Limits

The CWA became law on October 18, 1972, and required industrial dischargers to implement best practicable technology (BPT), best available technology economically achievable (BAT), or best conventional pollutant control technology (BCT) treatment. Pursuant to the CWA, the USEPA promulgated effluent limitation guidelines (ELGs) prescribing minimum treatment standards for discharges from various industries. The Water Board's NPDES permits impose technology-based effluent limits on industrial discharges based on the applicable ELGs. The ELGs are calculated based on the type of industrial process and, if applicable, the rate of production.

In 1982, the USEPA promulgated ELGs for the Petroleum Refining Point Source Category (40 CFR Section 419) for five refinery categories. The USEPA further amended these ELGs in 1985 and began re-evaluating them in 2017 (that effort is on-going). The ELGs prescribe technology-based effluent limits for petroleum refining wastewater calculated based on the total refinery throughput in barrels of crude oil per day, the type and configuration of refining processes, and the minimum acceptable treatment performance. The refinery ELGs cover biochemical oxygen demand, total suspended solids, chemical oxygen demand, oil and grease, phenolic compounds, ammonia as nitrogen, sulfide, total and hexavalent chromium, and pH. The Water Board implements the most stringent of the technology-based limits calculated under BPT, BAT, and BCT assumptions in its NPDES permits. The resulting technology-based limits are typically a daily maximum and monthly average in mass per day for each pollutant.

Water Quality-based Effluent Limits

Technology-based effluent limits are typically not adequate by themselves to meet water quality standards applicable to San Francisco Bay. The Water Board therefore imposes water quality-based effluent limits in the NPDES permits for any pollutants that might cause or contribute to violations of water quality standards. Water quality-based effluent limits are based on applicable water quality objectives or standards and characteristics of the discharge and receiving water. Water quality-based limits are usually a concentration-based daily maximum and monthly average for each pollutant, unless concentrations do not apply (such as for pH). The NPDES permits for refineries typically include water quality-based effluent limits for metals, dioxins, ammonia, pH, and aquatic toxicity.

TMDLs

San Francisco Bay and many of its tributaries do not meet water quality standards for several pollutants, and are therefore considered impaired under CWA Section 303(d). Industrial wastewater discharges to the Bay contribute to mercury and selenium impairment; discharges from a limestone mine and cement manufacturing facility to Permanente Creek cause or contribute to selenium and toxicity impairment. The Water Board developed Bay-wide TMDLs for the legacy pollutants mercury and polychlorinated biphenyls (PCBs) and for selenium in the North Bay to address all sources and pathways, including industrial discharges. The Water Board is also developing TMDLs for selenium and toxicity in Permanente Creek. The Water Board implements TMDLs either through watershed permits written specifically for that purpose or by imposing effluent limits based on TMDL wasteload allocations in existing individual NPDES permits.

Refinery discharges are subject to three TMDLs: mercury, PCBs, and selenium. The Water Board implements the Mercury and PCBs TMDLs through a watershed NPDES permit, last reissued in 2017 (see Watershed NPDES Permits, (**pages 13-14**). This Watershed Permit imposes effluent limits based on the Mercury and PCBs TMDL wasteload allocations for the five petroleum refineries, as well as other industrial facilities discharging directly to the Bay. The Water Board implements the North San Francisco Bay Selenium TMDL through provisions in the individual refinery NPDES permits, requiring industrial dischargers to monitor and report selenium loads and imposing performance-based mass effluent limits.

> Refinery wastewater treatment improvements and upgrades in the 1980s and 1990s resulted in significant improvements in effluent quality and reductions in discharge flow

Recent Findings

Flows

Refinery wastewater treatment improvements and upgrades in the 1980s and 1990s resulted in significant improvements in effluent quality and reductions in discharge flow. Refinery wastewater treatment plant operations and performance have been relatively consistent since then. In the mid-1980s there were six refineries in operation (including Pacific Refinery which closed in the mid-1990s) that discharged approximately 30 million gallons per day (MGD) of treated effluent. Water conservation efforts, as well as the use of reclaimed water in refinery processes, resulted in significant flow reductions at one of the refineries, which decreased its flows from nearly 20 to less than 8 MGD in the mid-2000s. The total flow from the five refineries in operation in the mid-2000s was 22 MGD. The total average daily flow from these refineries in 2017 and 2018 was very similar: 21 MGD.

Pollutants of Concern in Refinery Effluent

Refinery effluent is monitored for a wide variety of pollutants of concern in the Bay, but contributes more than 1% of the total loading of only one pollutant: selenium (**Figure 2**). For the North Bay – the portion of San Francisco Bay that is subject to the Selenium TMDL – the refineries contribute 11% of the load, which is higher than the loads from municipal wastewater and stormwater; the primary load into the Bay is from the Delta. Industrial effluent (of which refinery effluent contributes most of the loading) accounts for only a small proportion of the loading of the two main legacy pollutants of concern in the Bay: 0.4% for PCBs and 0.1% for mercury. As discussed in the article on municipal wastewater (**pages 2-21**), nitrogen is a pollutant of

Mercury

Delta (42%)

Urban Stormwater (26%)

Industry (0.1%)
Municipal Wastewater (0.6%)

Other Nonurban
Stormwater (6%)

Guadalupe River (20%)

Atmospheric Deposition (6%)

Nitrogen

Municipal
Wastewater (62%)

Refinery (1%)

Urban
Stormwater (15%)

Delta (22%)

PCBs

Stormwater (69%)

Industry (0.4%)

Municipal Wastewater (3%)

Delta (29%)

Selenium

Delta (77%)

Municipal Wastewater (2%)

Refinery (11%)

Urban and Nonurban
Stormwater (10%)

Figure 2. Mercury, PCBs, selenium, and nitrogen have been a focus of regulatory attention and the subject of inventories of loading to the Bay. Refinery effluent is monitored for a wide variety of pollutants of concern, but contributes more than 1% of the total loading of only one pollutant: selenium. For the North Bay – the portion of San Francisco Bay that is subject to the Selenium TMDL – the refineries contribute 11% of the load, which is higher than the loads from the municipal wastewater and stormwater pathways; the primary load is from the Delta.

FOOTNOTE: Pathway categories vary by pollutant because they were treated differently in the TMDLs.

FEATURE ARTICLE | INDUSTRIAL WASTEWATER

increasing concern in the Bay, but the nitrogen loading from refineries is also relatively small (1.3% of the total from the major pathways).

Pollutants in refinery effluent that cause toxicity to aquatic life were a concern prior to the addition of activated carbon treatment steps in the 1990s. In recent monitoring, effluent from the refineries is rarely found to cause toxicity in acute bioassays with rainbow trout, a sensitive test species.

Selenium

Since the refineries are a significant source of selenium to the North Bay, monitoring and studies to understand the impacts of refinery inputs have been a priority for the refineries, the Water Board, and the Regional Monitoring Program for Water Quality in San Francisco Bay (RMP). The refineries have made significant contributions to the science and understanding of selenium in the North Bay. They have worked collaboratively with the Water Board and provided funding for studies that contributed to the development of the North Bay Selenium TMDL. Refinery technical staff and their consultants serve on the RMP Selenium Workgroup and participated in designing a long-term monitoring program for selenium in the North Bay.

The selenium removal measures implemented by the refineries in the late 1990s yielded a substantial reduction in loading. Loads were greatly reduced by the mid-2000s and have remained at similar levels since that time (**Figure 3**). Total loads from the refineries ranged from 1,800-2,600 kg/yr in 1986–1992. The average load for 2009-2012 was 571 kg/yr. More recently, the average load for 2017-2018 was 526 kg/yr. The treatment upgrades also changed the proportions of the forms of selenium discharged, with a shift from selenite (Se IV) – the form of higher bioaccumulation concern – to the less bioavailable selenate (Se VI).

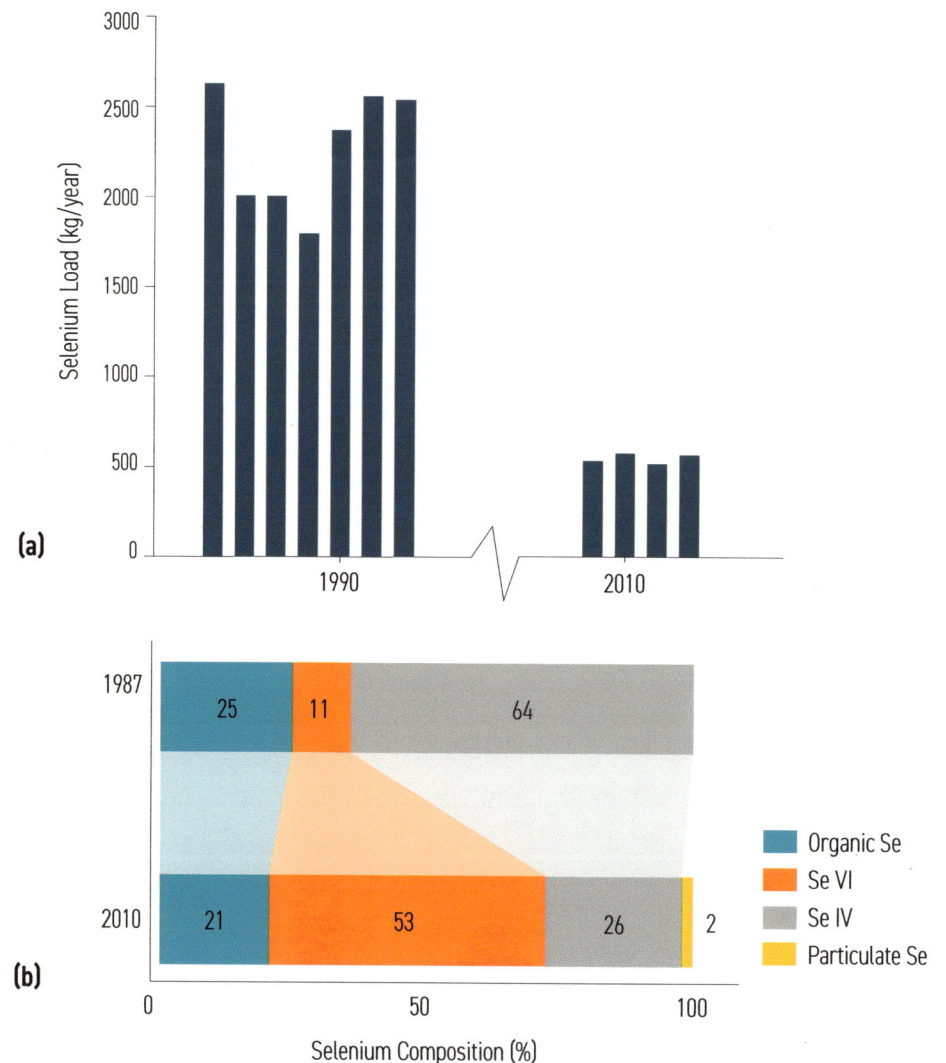

Figure 3. The selenium removal measures implemented by the refineries in the late 1990s yielded a substantial reduction in loading. a) Loads were greatly reduced by the mid-2000s and have remained at similar levels since that time. b) The treatment upgrades also changed the proportions of the forms of selenium discharged, with a shift from selenite (Se IV) – the form of higher bioaccumulation concern – to the less bioavailable selenate (Se VI).

FOOTNOTE: Based on average loads from five refineries. Particulate selenium was not measured prior to 2010. Adapted from Baginska (2015).

To support development of the TMDL, the refineries funded the North San Francisco Bay Selenium Characterization Study (Tetra Tech Inc. 2012). This study updated the information on selenium distribution and speciation under representative hydrologic conditions. The study measured selenium concentrations in the water column and on suspended particles, and evaluated selenium speciation and particulate selenium concentrations in refinery effluent and Bay water. Samples were also collected at small tributaries that flow directly into North Bay, which are distinct from the larger tributaries (the San Joaquin and Sacramento Rivers) that flow to the Bay through the Delta. The information from the study supported an update of a selenium model ("ECoS3") that was used in TMDL development (Tetra Tech Inc. 2015).

Under the guidance of the Selenium Workgroup, the RMP has conducted a series of studies in the last few years with the overall goal of establishing a cost-effective long-term monitoring plan for the North Bay to support implementation of the North Bay Selenium TMDL, which was approved in 2016. Monitoring is needed to track potential near-term and long-term changes. In the near term, changes in concentrations in the inflow from the San Joaquin River basin, refinery inputs, stormwater and tributary loads from the Bay margin, and overall Central Valley hydrological conditions (such as the extreme wet and dry periods that occurred between 2012 and 2017) may drive changes in concentrations in the North Bay. Other drivers, such as nutrient concentrations and algal levels, may also play a role, especially regarding selenium concentrations on particles. Over the longer term, selenium changes may occur due to modification of Delta flows and the mix of riverine sources because of the implementation of the WaterFix (Delta tunnels) project by the state of California. A primary monitoring goal is to identify leading indicators of change to allow prompt management response to signs of increasing impairment.

One recent study was done in coordination with the Sturgeon Derby, an annual fishing contest that focuses on white sturgeon in the North Bay

(Sun et al. 2019a). In a collaborative study involving the San Francisco Estuary Institute, the US Geological Survey, the US Fish and Wildlife Service, and Dr. Vince Palace of the International Institute for Sustainable Development (Winnipeg, Manitoba), selenium concentrations were measured in sturgeon tissue samples collected in 2015 through 2017. Multiple tissues were collected to develop laboratory methods for processing and analyzing selenium in muscle plug samples (a non-lethal method) and techniques for analyzing sturgeon fin rays. Otolith (ear bone) samples and tissues of greater toxicological interest (ovary and liver) were also collected. The study found a correlation between concentrations in tissue that can be collected non-lethally (muscle plug) and ovary and liver (**Figure 4**). This study complemented a parallel pilot effort to monitor selenium in sturgeon muscle plugs in the North Bay (**pages 88-89**).

Two other recent data synthesis studies conducted under the RMP to support development of a North Bay selenium monitoring design have focused on data synthesis. In the first (Chen et al. 2017), scientists from Tetra Tech evaluated the most recent changes in selenium in the North Bay and Delta through the analysis of observed data and modeling. One of the noteworthy findings was that selenium concentrations in the San Joaquin River at Vernalis (the southern boundary of the Delta) – one of the primary pathways for selenium into the Estuary - have decreased in recent years. Another key finding relevant to the refineries was that during average and wet years, loads from the Delta were the largest source of selenium to the Bay. During severe drought years, such as in 2014 and 2015, the Delta loads may be the same magnitude as the refinery loads.

The second study developed a robust monitoring plan for the North Bay (Grieb et al. 2018). The primary goals were to 1) identify leading indicators of change and 2) develop a program to monitor those key selenium indicators of water quality conditions in the North Bay. The key indicators include selenium concentrations in water, clams, and white sturgeon. Statistical analyses evaluated alternative monitoring strategies capable of detecting specified levels of change in each indicator. Overall, the findings showed that the implementation and continuation of long-term monitoring programs are required to identify changes from established baselines and to distinguish deviations from the effects of natural variability. In 2019 the RMP began a pilot phase of the recommended monitoring.

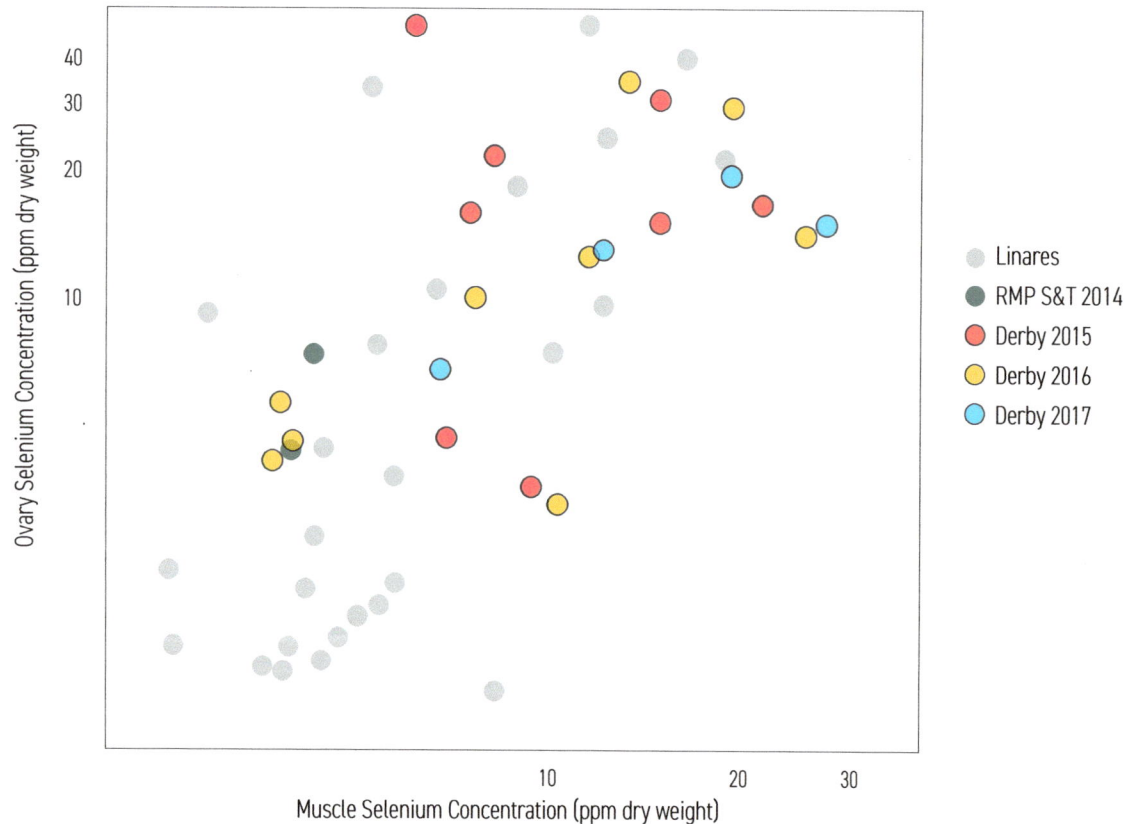

Legend:
- Linares
- RMP S&T 2014
- Derby 2015
- Derby 2016
- Derby 2017

X-axis: Muscle Selenium Concentration (ppm dry weight)
Y-axis: Ovary Selenium Concentration (ppm dry weight)

Figure 4. A 2015-2017 study done in coordination with the Sturgeon Derby, an annual fishing contest that focuses on white sturgeon in the North Bay, measured selenium concentrations in multiple tissues. The study found a correlation between concentrations in tissue that can be collected non-lethally (muscle plug) and ovary and liver. This study complemented a parallel pilot effort to monitor selenium in sturgeon muscle plugs in the North Bay (**pages 88-89**).

FOOTNOTE: Relationship between selenium concentrations measured in paired muscle and ovary samples. Data are shown on a log scale. Each point represents an individual fish. Data from the RMP Sturgeon Derby Study are shown as dots outlined in black. There was a statistically significant relationship between ovary and muscle selenium concentrations.

Future Work and Challenges

As they have in the past, Bay Area petroleum refineries will continue to strive to fully comply with existing and future regulatory requirements and do their part in protecting water quality in San Francisco Bay. The following changes and challenges are anticipated in the coming years.

USEPA Re-evaluation of Technology-based Effluent Limits

It is possible that the USEPA may issue requirements for petroleum refineries to enhance wastewater treatment technologies based upon revision of the national effluent limitations guidelines and standards required by CWA Section 304(b).

Water Sustainability

Water sustainability is a state-wide concern due to periodic drought conditions increasingly ascribed to climate change. Ensuring adequate supplies of drinking water and water for agriculture and industry will put significant emphasis on water conservation and reuse. Processing crude oil requires water, and all of the local refineries reuse water in various ways and evaluate opportunities for water recycling and reuse on an ongoing basis. For example, the Chevron Richmond Refinery has been able to build the capacity to meet over half of its water supply needs with recycled municipal wastewater from the East Bay Municipal Utility District – the water is used in cooling towers and as makeup water for boilers. As another example, Phillips 66 has plans to recycle the vast majority of their effluent and re-use it in the refining process. Several refineries are also able to use treated effluent in fire protection and firefighting.

Sea Level Rise and Climate Change

Sea level rise (SLR) and climate change are of concern to facilities with considerable operational infrastructure located on or near the shoreline of San Francisco Bay. Municipal wastewater treatment plants, railroads, industrial facilities, and petroleum refineries may be affected in the future.

California Executive Order S-13-08, issued on 14 November 2008, recognized the impact that SLR may have on coastal development, existing facilities, and infrastructure in California, and directed state agencies to plan for SLR and coastal impacts. The Executive Order also requested the National Research Council (NRC) to issue a report on SLR to advise California on planning efforts. A State of California Sea-Level Rise Interim Guidance Document was released from the Sea-Level Rise Task Force of the Coastal and Ocean Working Group of the California Climate Action Team (CO-CAT) in 2010 and the final report, Sea-Level Rise for the Coasts of California, Oregon, and Washington, was released from NRC in June 2012. The State of California Sea-Level Rise Guidance Document was updated by CO-CAT member agencies in 2018. These reports advise California state agencies on how California communities should plan for SLR.

Regional and local agencies that are reviewing development and construction permit applications regarding public infrastructure and private development are taking a proactive approach in response to the NRC and CO-CAT guidance. In 2010, the San Francisco Bay Conservation and Development Commission (BCDC) partnered with the NRC to evaluate impacts of SLR specifically on communities surrounding San Francisco Bay. Recently, BCDC has established the "Adapting to Rising Tides Group," which is a collaborative planning effort to help Bay Area communities adapt to SLR and storm event flooding. These organizations and others will continue to issue new guidance as new scientific information is developed and released. It is important to note that no regulatory policy related to SLR has been formally adopted by any federal, state, or local agencies to date.

On-going elements of infrastructure management will be to:

- define local and site-specific parameters that define the basis for SLR at the shoreline adjacent to a given facility;

- conduct vulnerability analyses to consider the direct and consequential effects of SLR; and

- develop adaptive management strategies to manage and mitigate the undesirable and consequential effects of SLR on facilities. §

STORMWATER

Lester McKee and **Jay Davis**
San Francisco Estuary Institute

Richard Looker and **Tom Mumley**
San Francisco Bay
Regional Water Quality Control Board

Chris Sommers
EOA Inc.

HIGHLIGHTS

▶ Rain falling on land surfaces in the San Francisco Bay watershed produces stormwater runoff that carries pollutants, mostly untreated, into storm drains, creeks, and ultimately, the Bay

▶ Although small local tributaries contribute only 6% of the fresh water entering the Bay, they contribute a disproportionately large percentage of the pollutant load

▶ Urban stormwater is the largest pathway for many pollutants of concern in the Bay, including PCBs, dioxins, PAHs, many trace metals and pesticides, microplastics, and potentially other types of emerging contaminants

▶ The Municipal Regional Stormwater Permit (MRP) regulates stormwater discharges from most of the municipalities in the Bay Area

▶ Since 2015, stormwater monitoring has focused on screening a large number of watersheds for PCBs and mercury, identifying those with relatively high concentrations that signal a greater potential for cost-effective management

▶ Priority future focus areas include integration of green stormwater infrastructure into the urban portion of watersheds, tracking trends in legacy pollutants, and investigating emerging contaminants

◀ **Stormwater outfall, Berkeley** (Shira Bezalel, SFEI)

Stormwater 101

The rainwater that falls on the regional landscape flows down hillsides and through the urbanized lowlands, travelling toward San Francisco Bay in storm drains, creeks, and channels. This water is referred to as stormwater. Stormwater is one of the most important pathways by which pollutants enter San Francisco Bay. The sources of pollutants in stormwater are the uses of chemicals and other activities that result in pollution of land surfaces. Patches of polluted soil and other surfaces distributed throughout urban and nonurban areas of the Bay watershed are the source areas that contaminate stormwater. When rain falls onto these surfaces, pollutants are mobilized and transported into storm drains, creeks, and ultimately, the Bay (**Figure 1** and **pages 40-41**). While stormwater flows and pollutant loading also impact creeks, the focus of this article is on stormwater as a pathway for pollutant input into the Bay.

Stormwater runoff that reaches the Bay drains from the Bay's watershed, which includes flows from the Central Valley and areas draining into local tributaries in the nine counties fringing the Bay. Given its large area, the Central Valley contributes the vast majority, roughly 89%, of the fresh water to the Bay. Only 6% of freshwater inputs are attributable to the smaller local tributaries, with the balance from rainfall directly onto the Bay surface (part of the "atmospheric deposition" pathway) and from treated wastewater (**Figure 2**). Small tributaries contribute a relatively small amount of fresh water to the Bay, but a disproportionately large load of pollutants.

Stormwater runoff does not occur equally from all land areas in the watershed. More runoff comes from areas that receive more rainfall or have more impervious surfaces such as roads, roofs, and parking lots. An estimated 59% of the runoff that reaches small tributaries around the Bay comes from areas with urban land uses, with the remainder from nonurban areas (open space and agriculture) (**Figure 3**). Stormwater originating from urban areas adjacent to the Bay often has pollutant concentrations that are tens to hundreds of times greater than those in flows from the Central Valley that enter the Bay from the Delta. For example, concentrations of PCBs in outflows from the Delta average just 0.34 ng/L, whereas the average from the tributaries that drain predominantly older urban and industrial areas in the nine counties around the Bay is about 16 ng/L (50 times greater). Even though flows in the tributaries are smaller than those from the Delta, high pollutant concentrations still result in relatively large loads to the Bay.

Urban stormwater is the largest pathway to the Bay for many pollutants of concern, including PCBs, dioxins, PAHs, many trace metals and pesticides, microplastics, and likely other emerging contaminants. Fuel, oil wastes, combustion products, and wear debris (trace metals, rubber, and plastics) from vehicles can accumulate on impervious surfaces (e.g., pavements and parking lots). Other urban stormwater pollutants are chemicals used in the urban environment, such as pesticides and herbicides that are applied around buildings and along roads, and nutrient-containing fertilizers used in landscaping. PCBs and mercury are important legacy pollutants that reached peak use in the 1960s and 1970s. Most often, sources and source areas for mercury and PCBs include properties in older industrial and commercial areas where they were commonly used or produced. Pollutants may be associated with contaminated soils, leaking electrical equipment, building materials, waste products and debris, paints and sealants, and aging machinery. Pollutants in all of these source areas can be washed downstream during winter storms and enter the Bay, mainly without treatment, via the urban stormwater drainage system.

Pollutant loads from urban stormwater vary tremendously by season. Stormwater loads follow episodic winter storm runoff patterns. Rainfall varies considerably from month-to-month and year-to-year as illustrated by a long-term rainfall dataset for San Jose (**Figure 4**). Larger stormwater loads are transported to the Bay during wetter years, and larger storms during any year tend to transport larger loads.

PIEDMONT

OAKLAND

ALAMEDA

Figure 1. An example stormwater collection system. The City of Oakland has 402 miles of storm drain pipes, more than 7,500 storm inlets, 137 weirs (areas where natural water courses enter the storm drain infrastructure), and five primary trash collection units (CDS), in addition to City creeks.

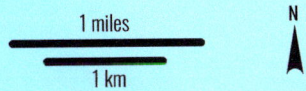

1 miles

1 km

N

Stormwater: Sources, Pathways, Loading

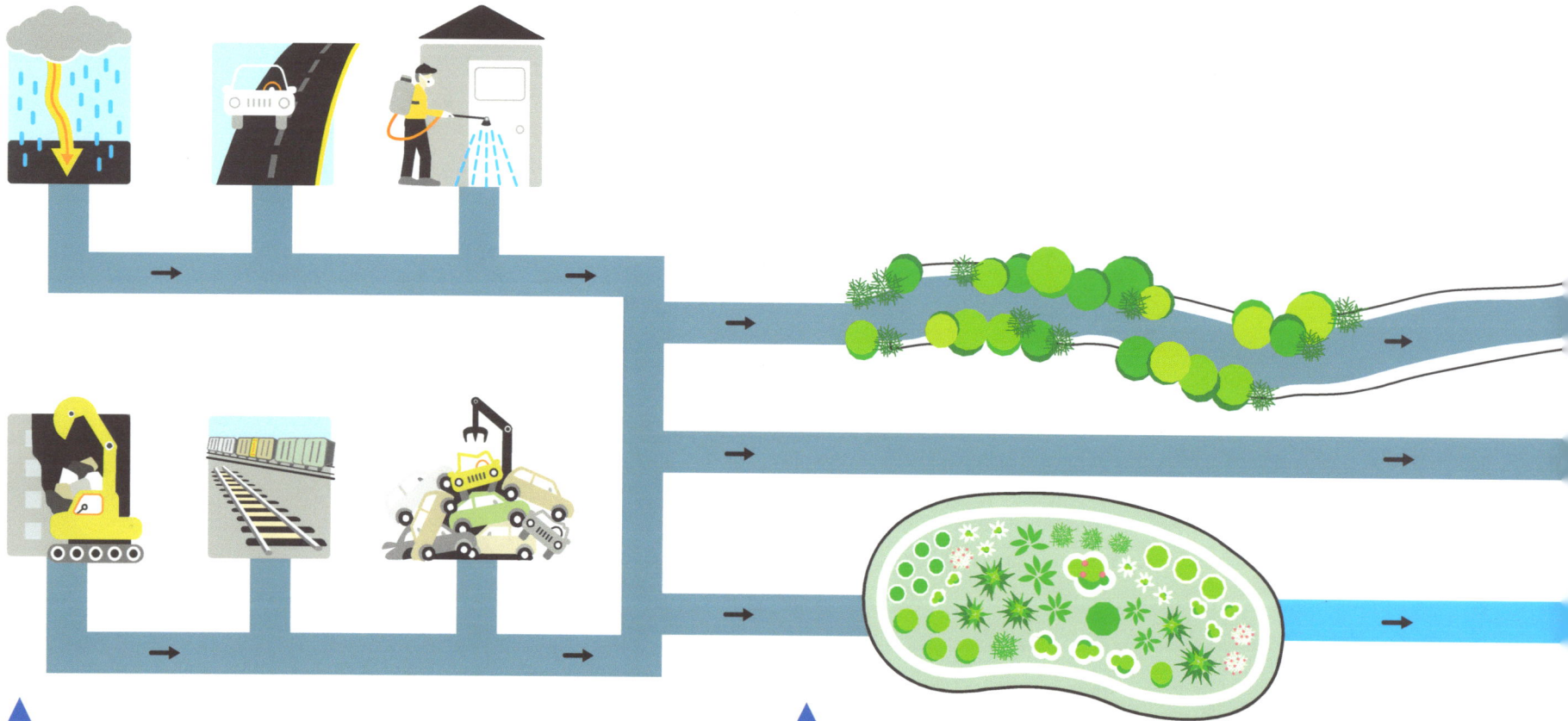

The **SOURCES** of pollutants in stormwater are the activities that result in pollution of land surfaces. Some of the most important sources and source areas are vehicle use, pesticide application, legacy contaminated areas (railroads, former industrial sites), demolition and construction, and active contaminated areas (junkyards, homeless encampments, illegal discharge and dumping, active industrial sites). Atmospheric deposition is an important pathway by which pollutants are deposited widely across the land surface and make their way into stormwater.

Stormwater is a **PATHWAY** by which pollutants from source areas in the watershed are transported through creeks and storm drains, downstream to the Bay. Most of this stormwater is untreated, but increasing amounts are routed through green stormwater infrastructure (such as rain gardens, bioswales, and tree wells) that filters out many pollutants. Bay Area cities and counties are developing long-term regional plans that transition from traditional gray infrastructure to green infrastructure.

A substantial portion of the **LOADING** of many pollutants of concern in the Bay is attributable to stormwater. Urban stormwater is the largest pathway for PCBs, dioxins, PAHs, many trace metals and pesticides, microplastics, and likely other emerging contaminants. Urban stormwater is also a pathway with a high potential and need for reduction of PCBs and mercury, two of the pollutants of greatest concern in the Bay.

Illustration by Linda Wanczyk (lindawanczyk.com)

Figure 2. Small tributaries contribute a relatively small amount of fresh water to the Bay, but a disproportionately large load of pollutants. The Central Valley contributes the vast majority, roughly 89%, of the fresh water to the Bay. Only 6% of freshwater inputs are attributable to the smaller local tributaries, with the balance from rainfall directly onto the Bay surface (atmospheric deposition) and from treated wastewater.

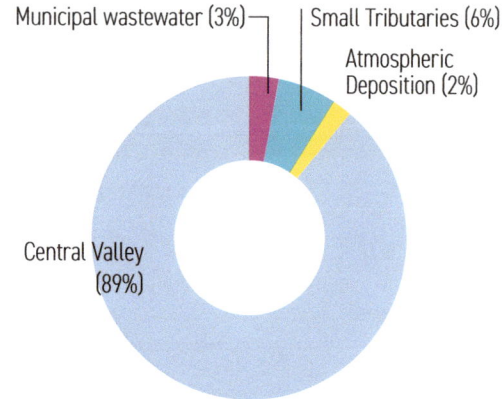

Municipal wastewater (3%)
Small Tributaries (6%)
Atmospheric Deposition (2%)
Central Valley (89%)

Figure 3. Urban areas in the watersheds of small tributaries account for a disproportionate amount of the flow, and most of the pollutant load, from this pathway. More runoff comes from areas that have more impervious surfaces such as roads, roofs, and parking lots. An estimated 59% of the runoff that reaches small tributaries around the Bay comes from areas with urban land uses, with the remainder from nonurban areas (open space and agriculture), even though urban land uses make up a smaller percentage (41%) of the overall land area.

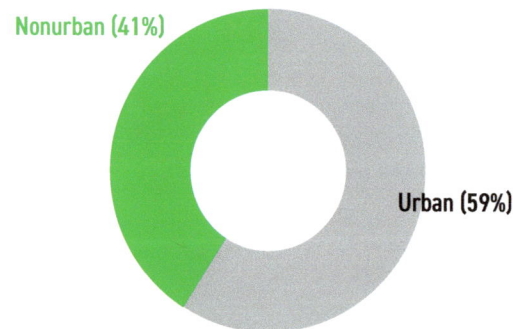

Nonurban (41%)
Urban (59%)

For example, based on long-term measurements of suspended sediment loads in the Guadalupe River carried out by the USGS in cooperation with Valley Water (formerly Santa Clara Valley Water District), 46% of the total suspended sediment load over 16 years (water years 2003-2018) was transported by a series of very large storms that hit the South Bay in January and February 2017. Similarly, the loading of mercury and other pollutants are very episodic. During a single January 2017 three-peak storm series, an estimated 70 kg of mercury (mostly from non-urban areas) was transported by the Guadalupe River to the Bay in just seven days, nearly equivalent to the long-term annual average load from this watershed (McKee et al., 2018). Loads of PCBs and other urban pollutants are also very episodic.

Suspended sediment, which serves as a transport mechanism for many particle-associated pollutants of concern, is also delivered via stormwater flows. The amount of sediment reaching the Bay has declined significantly over the last century because of changes humans have made to the broader Bay watershed, which includes the Central Valley. The Central Valley used to deliver the largest supply of sediment to the Bay, but over time this has gradually changed. Dams built for water supply have curtailed sediment transport from 50% of the Sierra Nevada area of the watershed, and significantly reduced peak winter flows. Suspended sediment concentrations in Delta outflow now average just 35 mg/L – about one twentieth of the average concentration of the small local tributaries around the Bay. At present, about 70% of the annual average of two million metric tonnes of suspended sediment entering the Bay is from local tributaries.

While urban stormwater from local tributaries is a major pathway for many pollutants, stormwater runoff from the Central Valley is the predominant pathway for selenium and mercury to the Bay. Although selenium concentrations are slightly higher in Bay Area urban

San Jose Annual Rainfall

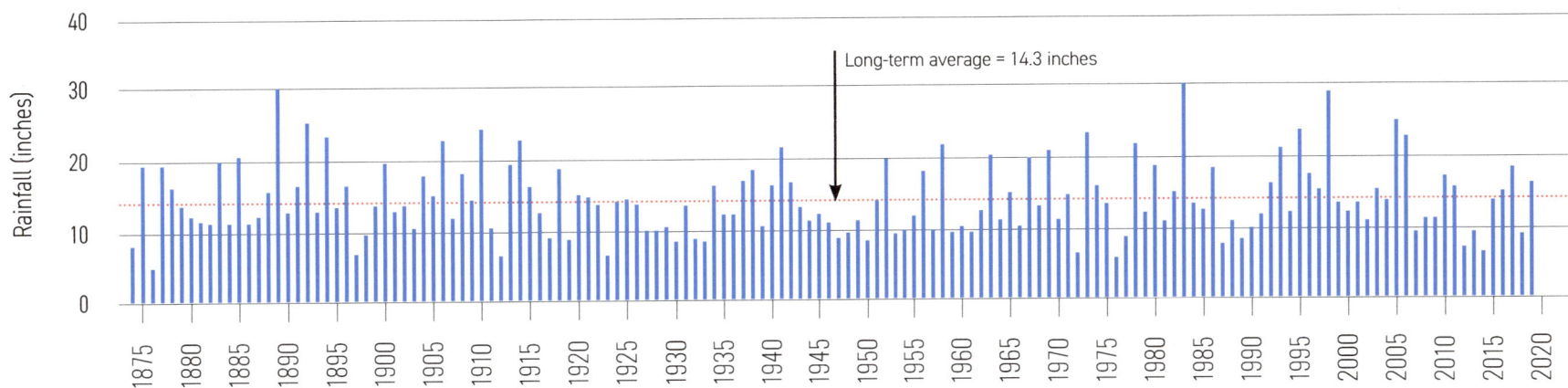

Long-term average = 14.3 inches

Figure 4. Following patterns in rainfall, pollutant loads from urban stormwater vary tremendously by year and season. Year-to-year variation in rainfall is illustrated by a long-term dataset for San Jose that goes back to 1874. Larger stormwater loads are transported to the Bay during wetter years, and larger storms during any years tend to transport larger loads.

FOOTNOTE: Data are for climate years (July 1 to June 30 with the year corresponding to the end date). Source: Jan Null, Golden Gate Weather Services.

stormwater, agricultural practices exacerbate the release of naturally occurring selenium from soils in the western Central Valley. This release, coupled with the huge volume of stormwater transported from the Central Valley via the Sacramento and San Joaquin rivers, results in a greater load of selenium via the Delta than from small local tributaries. Mercury was mined in Bay Area watersheds and used extensively during the Gold Rush era at over 2,000 mine sites in the Sierra Nevada foothills. The use of mercury for gold extraction left behind a legacy of watershed contamination and loading that is still impacting the Bay today. Due to the large runoff volume from the Delta, the Delta load of mercury still exceeds other large pathways, including stormwater from local tributaries, in spite of the impacts of atmospheric deposition onto the watersheds and runoff from the historic New Almaden Mining District that drains to the South Bay. Heavy use of fertilizers on Central Valley crops and nutrient discharges into Central Valley tributaries from urban wastewater sources lead to a larger load of nutrients than from urban stormwater from small local tributaries. This load, however, is still far exceeded by the nutrient loads to the Bay from treated municipal wastewater (**pages 2-21**).

Regulatory Framework

Stormwater Management via the NPDES Program

Given the episodic nature of pollutant transport in stormwater and the difficulty of locating elusive sources and source areas (as described in the previous section), managing stormwater quality is challenging. Efforts to regulate stormwater quality in the Bay Area began with an amendment to the Water Quality Control Plan for the San Francisco Bay Basin (the "Basin Plan") in 1986. That plan established water quality objectives for several metals and other toxic pollutants and recognized that the objectives could not be attained without managing stormwater contributions to the Bay. Soon after, the 1987 amendments to the federal Clean Water Act and the ensuing regulations promulgated by the US Environmental Protection Agency (USEPA) in 1990 established the National Pollutant Discharge Elimination System (NPDES) permit program for stormwater, 18 years after the NDPES permit program for municipal wastewater was established.

Similar to the NPDES permit program for municipal wastewater in California (**page 11**), the USEPA delegates authority to the San Francisco Bay Regional Water Quality Control Board (Water Board) to issue permits to Bay Area cities and towns, counties, and flood control districts that require the management and monitoring of stormwater. Rather than issuing separate permits to individual local agencies, the Water Board and municipalities agreed in the late 1980s that permits covering county-wide areas would be more efficient and manageable than issuing individual permits to each public agency. The first NPDES stormwater permit in the Bay Area was issued to public agencies in Santa Clara County in 1990. Similar permits went into effect in Alameda County in 1991; Contra Costa and San Mateo Counties in 1993; the cities of Fairfield and Suisun City in 1995; and Vallejo in 1998. In 2003 the State Water Board issued a permit for storm drain discharges from small municipalities in the North Bay and the portion of San Francisco that does not drain to its combined sewer system. These permits primarily called for implementation of self-determined management plans covering actions in categorical program areas. This approach is described in the Urban Runoff Management Program section of the Basin Plan.

The permits required municipalities to implement a suite of controls to reduce pollutants to the maximum extent practicable. The requirements covered: operation and maintenance of storm drain systems, streets and roads, parks, and other public facilities; new development and redevelopment; commercial and industrial facilities; construction sites; and outreach and education. Permits also required municipalities to seek out, eliminate, and prohibit any illicit connections and discharges of non-stormwater flows to their storm drain systems. More detail was added to the stormwater NPDES permits when they were reissued in later years, but the management plan approach continued, which allowed flexibility and adaptation as municipal efforts evolved and matured. This approach was largely successful, as illustrated by the numerous national and state awards received by Bay Area stormwater programs over the past 25 years.

Despite these successes, however, in the mid-2000s the Water Board and municipalities began to recognize the potential benefits of a single regional permit that could address the growing attention to water bodies listed as impaired under Section 303(d) of the Clean Water Act and the resulting Total Maximum Daily Load (TMDL) control plans, the public scrutiny of management plan-based permits, and the desire for more consistency and efficiency. In 2009 the Water Board consolidated county-based NDPES permits and adopted a new Municipal Regional Stormwater Permit (MRP) that covers most of the public agencies in the Bay Area. The consolidated MRP has addressed many of the concerns and raised the effectiveness of stormwater management in the Bay Area to a higher level. The MRP was reissued in 2015 and another update is planned for 2021.

In contrast to permits previously issued on a county-wide basis or to individual cities and counties, each iteration of the MRP has included water quality-based requirements for specific pollutants to address TMDLs and other pollutants of concern. These pollutant-specific requirements started with copper and nickel, and eventually included pesticides, mercury, PCBs, bacteria, and trash. Permit monitoring requirements to characterize the presence and loading of pollutants, identify sources of pollutants, and evaluate the effectiveness of control measures also have a nexus with RMP management questions. Accordingly, throughout its history, the RMP has played a key role in generating data and information to meet permit requirements and adaptive improvements.

Informing Bay Area Stormwater Management via the RMP

Over the past 25 years, the RMP has provided vital information to assist with the effective management of stormwater in the Bay Area. Interest in urban stormwater was a major reason for establishing the RMP Sources, Pathways, and Loadings Workgroup (SPLWG) in 1999, and this Workgroup has been central to the RMP's role in providing data and information relevant to stormwater regulation. The SPLWG has shaped the conceptual understanding of how contaminants are transported to the Bay and developed contaminant loading estimates based on a variety of computational methods. An important early milestone was the first SPLWG Technical Report (Davis et al. 2001) that estimated loads of PCBs, PAHs, pesticides, and a variety of metals. The information needs identified in this report became a roadmap for work conducted through the RMP workgroups and special studies over the subsequent two decades.

Beginning in the late 1990s, the Water Board relied on RMP monitoring data and technical analysis to focus attention on copper and nickel as key pollutants of concern and to achieve acceptance of higher (yet still protective) site-specific water quality objectives (SSOs) for copper. The robust historical RMP dataset for these metals not only provided a foundation for the Water Board to establish the copper SSOs, but the ongoing RMP status and trend monitoring dataset also gives the Water Board and its stakeholders confidence that water quality is being protected through the pollution prevention requirements in wastewater and stormwater NPDES permits.

The RMP has played an ongoing essential role in advancing understanding of pollutant contributions from stormwater to the Bay, and the need for enhanced stormwater management for some pollutants. For example, along with data collected by Bay Area cities, counties, and flood control districts, the RMP helped establish the need for enhanced actions to manage PCBs and mercury contributions from stormwater in the Bay Area. TMDLs adopted by the Water Board for PCBs and mercury, including wasteload allocations for urban stormwater, were established using these data, despite some information and data limitations. The Water Board is able to regulate in the face of such uncertainty in large part because the RMP provides an effective means to fill data and information gaps so TMDLs can be adapted if needed. RMP studies indicated that urban stormwater was the dominant pathway for PCBs to reach the Bay, which translated into a 90% reduction requirement for urban stormwater in the TMDL. Additionally, the availability of the RMP to conduct additional

> The RMP has played an ongoing essential role in advancing understanding of pollutant contributions from stormwater to the Bay, and the need for enhanced stormwater management

studies on PCBs helped justify a phased implementation plan in the MRP that could be adapted based on further special studies of PCBs in the Bay and its urbanized watersheds.

RMP studies also showed that urban stormwater is a pathway for mercury to reach the Bay, with the legacy of mercury mining in the Guadalupe River watershed, use of refined mercury in gold mining in the Central Valley, and manufacturing processes and products contributing to the mercury in Bay fish in different areas. The importance of mining mercury had been a working assumption and was confirmed by a post-mercury TMDL special study using mercury isotopes. Additionally, RMP studies recently generated an improved estimate of the mercury load from urban stormwater that was 25% percent lower than that used to develop the Mercury TMDL. These studies have continued to improve our understanding of the importance of stormwater in the protection of Bay water quality.

With the inclusion of pollutant-specific requirements in the MRP and the implementation of enhanced stormwater management actions, an even closer collaboration between stormwater programs and the RMP has occurred through the Small Tributaries Loading Strategy (STLS) Team, a small group of stormwater program representatives, Water Board staff, and RMP staff that are collaborating on a number of high priority efforts. The STLS Team has developed conceptual models of source release processes for stormwater pollutants of concern and applied these conceptual models to identify watershed areas with elevated concentrations of legacy pollutants, such as PCBs and mercury. These

> Preliminary results from a 2016 RMP special study suggest that stormwater has the potential to contain significant levels of contaminants of emerging concern

areas may be candidates for enhanced stormwater management actions via the MRP. Additionally, the RMP recently used this monitoring information to calibrate a GIS-based numerical model of watershed loads of PCBs and mercury. The outputs of this model are currently being used by MRP stormwater programs as they develop control measure implementation plans for PCBs and mercury as part of the third generation of the MRP. The information obtained through these RMP initiatives continues to inform control measure implementation by MRP permittees and provides a basis to refine urban stormwater wasteload allocations as the PCBs and Mercury TMDLs are adapted.

Although the stormwater focus over the past decade has been on legacy pollutants, the RMP has identified a growing list of contaminants of emerging concern (CECs) that may have a high probability of at least low-level effects on Bay aquatic life ("moderate concern CECs" - **page 76**). The RMP Emerging Contaminants Workgroup (ECWG) has begun investigating pathways for emerging contaminants, and preliminary results from a 2016 RMP special study suggest that stormwater has the potential to contain significant levels of CECs. The ECWG, in partnership with the SPLWG and STLS, is currently following up on this initial finding with a three-year screening study of emerging contaminants in urban stormwater. Stormwater samples collected from urban watersheds are being analyzed for several classes of moderate concern CECs, such as urban pesticides, as well as compounds from vehicle tires, some of which may be implicated in the Puget Sound area as possible causes of toxicity to coho salmon. Additional information on CEC monitoring is discussed in the "Future Work, Directions, and Challenges" section.

Recent Findings

Stormwater Loading

Given the importance of stormwater runoff to loading of pollutants of concern to the Bay, considerable effort has been made over the past 25 years to quantify pollutant concentrations and loads in urban stormwater and small tributaries. It is currently estimated that, on average, about 20 kg of PCBs and 120 kg of mercury enter the Bay from urban runoff annually (**Figure 5**). The Guadalupe River watershed, with its mercury mining legacy, discharges another 92 kg of mercury per year. In contrast, the best current selenium loading estimates suggest that 77% (4,070 kg/yr) of the total comes from the Central Valley, compared to 520 kg in stormwater from North Bay small tributaries. It is estimated that urban stormwater from Bay Area tributaries also accounts for 15% of the average daily nitrogen load (10,820 kg/d), while wastewater facilities discharge 62%, (45,000 kg/d).

Trends in Pollutant Concentrations and Loads

Little work has been done so far to investigate trends in loadings from stormwater, although the SPLWG and STLS Team are beginning to implement a new trends monitoring and modeling strategy. No evidence currently exists to suggest a loading or concentration trend for PCBs and mercury based on direct measurements of stormwater. However, cores collected in 2006 from wetlands around the Bay (Yee et al. 2011) suggest that concentrations of PCBs and mercury have declined since periods of peak loading in the 1960s. A wetland core from Damon Slough at the edge of San Leandro Bay showed some of the greatest changes, with a five-fold difference in mercury between the buried peak concentrations and the near-surface layer. PCBs from that location similarly showed a near three-fold decrease from their peak. However, continued decreases in PCB concentration in San Leandro Bay are less evident. Although the site of the Damon Slough core was not resampled, surface sediment in the adjacent channel and other San Leandro Bay channel and subtidal

sites nearby suggest no significant decline between samples taken in 1998 (Daum et al., 2000) and samples taken at many of the same sites in 2016 (Davis et al., 2017). Thus progress appears to have stalled or greatly slowed in recent decades.

A recent report evaluated PCB trends in the Guadalupe River using a statistical model that considered climatic variation (Melwani et al., 2018). The authors assessed the power of various sampling designs to potentially observe a 25% load reduction over a 20-year period. The model did not suggest a significant linear inter-annual trend in PCB loads for the period 2003-2014 after accounting for climatic variability. The study found that if four grab samples per storm were collected during four storms every second year going forward, it would be possible to detect a trend of greater than 25% decline over 20 years. Although other watersheds will have different runoff and pollutant generation processes, this design provides a useful starting point for future trends monitoring in other watersheds.

The Guadalupe River has also been a focus of mercury trend monitoring. In this watershed, it is estimated that 86% of the mercury is derived from nonurban sources within the historic mining areas in the upper watershed, with the balance from the urban area downstream. Based on comparisons between the data collected by the RMP during a January 2017 storm and previous data collected during similar storms in December 2002, the characteristics of mercury transport in the system have not changed. Stormwater and sediment particle concentrations; the proportion of transport in dissolved phase; the relationship between instantaneous flow, suspended sediment loads, and instantaneous mercury loads; and the mechanisms of transport (rainfall intensity and flow sources) appear to remain unchanged over the 14-year period despite considerable effort to clean up some of the primary source areas. These sobering results illustrate the formidable challenge of managing legacy pollutants for which a large supply remains in the watershed, especially in the face of extreme events that exceed the design criteria of corrective measures.

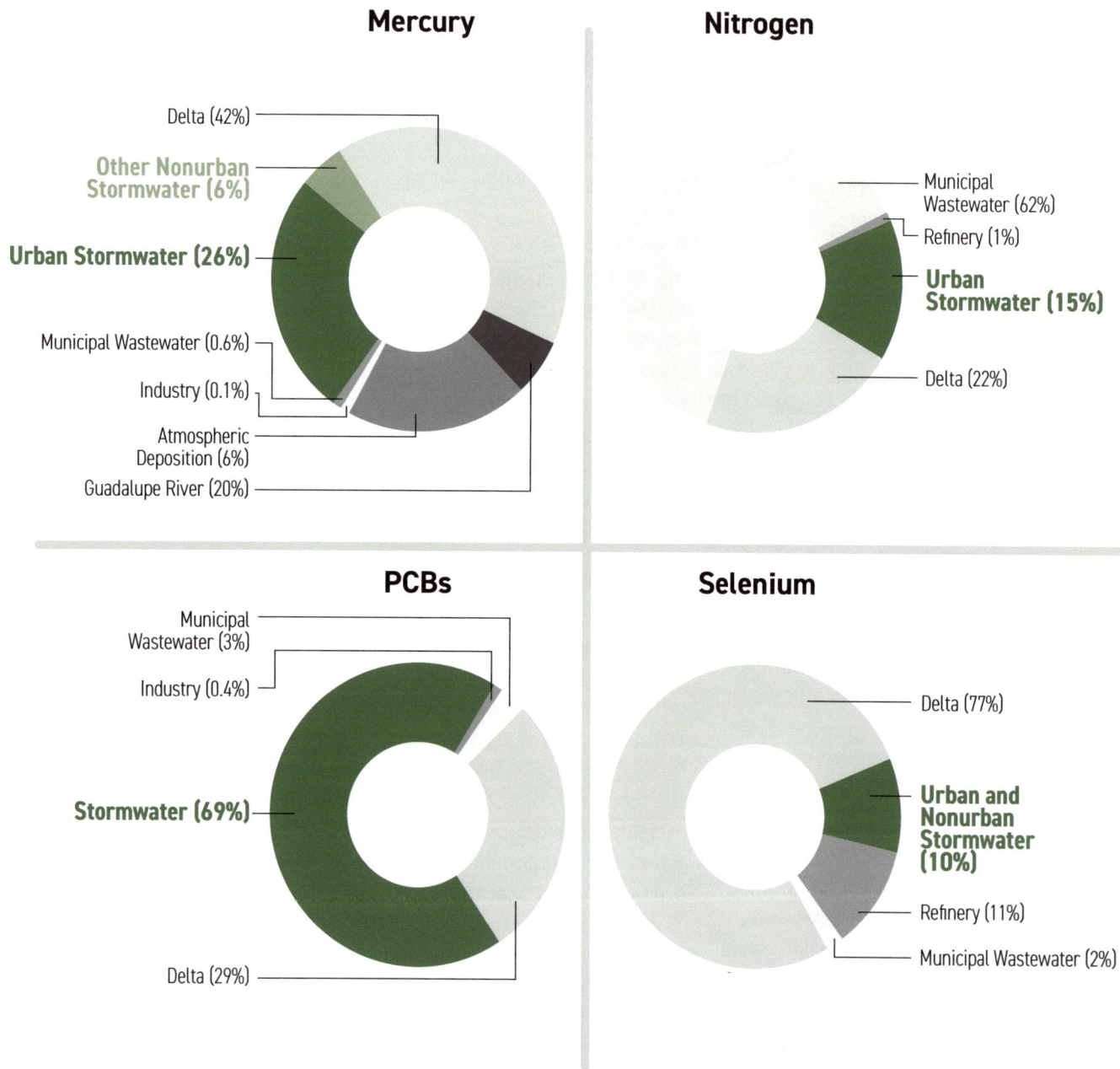

Mercury

- Delta (42%)
- Other Nonurban Stormwater (6%)
- Urban Stormwater (26%)
- Municipal Wastewater (0.6%)
- Industry (0.1%)
- Atmospheric Deposition (6%)
- Guadalupe River (20%)

Nitrogen

- Municipal Wastewater (62%)
- Refinery (1%)
- Urban Stormwater (15%)
- Delta (22%)

PCBs

- Municipal Wastewater (3%)
- Industry (0.4%)
- Stormwater (69%)
- Delta (29%)

Selenium

- Delta (77%)
- Urban and Nonurban Stormwater (10%)
- Refinery (11%)
- Municipal Wastewater (2%)

Figure 5. Mercury, PCBs, selenium, and nitrogen have been a focus of regulatory attention and the subject of inventories of loading to the Bay. Stormwater from local small tributaries is a major pathway for PCBs and mercury: the two pollutants causing the greatest impairment of Bay water quality. Local stormwater also accounts for significant loads of nitrogen and selenium. Mercury loads from small tributaries come from a combination of urban stormwater and nonurban stormwater from the historic New Almaden Mining District in the Guadalupe River watershed. Loads from the Delta, another form of stormwater input, are also significant.

FOOTNOTE: Pathway categories vary by pollutant because they were treated differently in the TMDLs.

Spatial Patterns of Pollutants in Watersheds

Despite limited results from trend analysis, an ever-growing understanding of the spatial patterns of PCBs, mercury, and other pollutants in urban stormwater is being built through RMP and stormwater program efforts. Since 2015, RMP stormwater monitoring has focused on characterizing single-storm average concentrations in urban stormwater from older industrial areas where higher concentrations and loads are expected. The most recent report (Gilbreath et al., 2019) provides a summary of PCB and mercury concentrations at 83 locations (**Figure 6**). About 33% of these sites have estimated PCB sediment particle concentrations greater than 200 ng/g and 40% have mercury concentrations above 500 ng/g (levels used as thresholds for considering source investigations). The three sites with the highest estimated PCB concentrations were Pulgas Pump Station South, Industrial Road Ditch in San Carlos, and Line 12H at Coliseum Way in Oakland. As shown in **Figure 6,** however, there are many other sites exhibiting high concentrations that may have a higher potential for cost-effective management.

One interesting use of these data has been to support loading estimates for a more detailed analysis of three "priority margin units." Mass budgets developed for PCBs in Emeryville Crescent, San Leandro Bay, and Steinberger Slough suggest that load reductions should lead to local improvements in these margin units. For example, if loads to the Emeryville Crescent were reduced to zero, it is estimated that sediment PCB concentrations in this area could reach the Central Bay ambient average in roughly 10 years, potentially leading to greatly lowered concentrations in local benthic animals and fish.

> The estimated microplastic load from stormwater is approximately 300 times higher than that of municipal wastewater

Contaminants of Emerging Concern in Stormwater

During the winters of 2016-2017, 2017-2018, and 2018-2019, considerable effort was put into sampling for CECs in stormwater. Microparticles, including microplastics, were detected in stormwater from all 12 small tributaries sampled, at concentrations between 1.3 and 30 microparticles per liter. Although there is considerable uncertainty, a first order estimate suggests approximately 7 trillion microplastic particles enter the Bay annually via stormwater, based on the output of the Regional Watershed Spreadsheet Model and the chemical characterization of a subset of the microparticles using spectroscopy. The estimated microplastic load from stormwater is approximately 300 times higher than that of municipal wastewater. The complete results of this study will be released in October 2019 (Sutton et al. 2019).

Stormwater sampling for other CECs began in the winter of 2018-2019 at eight sites and will continue in the winter of 2019-2020. The analytes include PFAS, ethoxylated surfactants, organophosphate esters, bisphenols, and a suite of stormwater CECs that are related to urban roadways and pesticides. The results of the first winter of sampling are still pending.

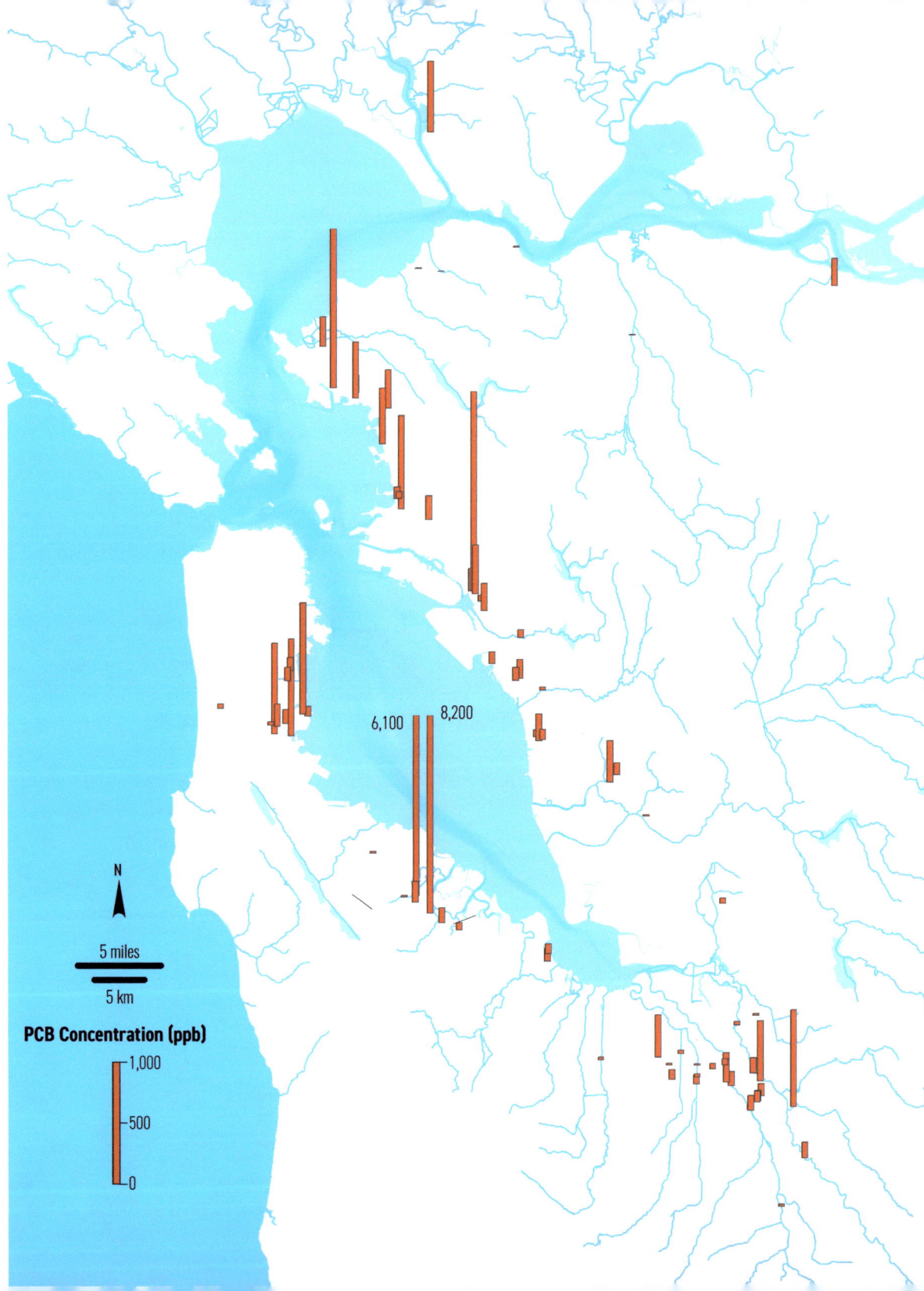

Figure 6. Urban stormwater is the largest pathway of PCB loads to the Bay. As a major element of the RMP in the last few years, concentrations of PCBs and mercury on suspended sediment particles from a wide range of watersheds are being measured as an index of the degree of watershed contamination and potential for effective management action. Stormwater samples from Pulgas Creek Pump Station North and South, Industrial Road Ditch, Gull Drive Storm Drain, and Outfall to Colma Creek in San Mateo County; Santa Fe Channel in Contra Costa County; Line 12H at Coliseum Way, Outfall at Gilman Street and Ettie Street Pump Station in Alameda County; and Outfall to Lower Silver Creek in Santa Clara County had the highest concentrations of PCBs on suspended sediment particles measured to date.

FOOTNOTE: Bars represent the average PCB concentration on suspended particles measured during a storm at each location. Note that the bars for the two stations with the highest concentrations (8,200 ppb and 6,100 ppb) are truncated. Data from Gilbreath et al. (2019).

N

5 miles
5 km

PCB Concentration (ppb)

1,000

500

0

6,100   8,200

FUTURE WORK, DIRECTIONS, AND CHALLENGES

The importance of urban stormwater pollutant loads to the Bay via small tributaries has led to an increased focus on controlling this pathway. Management actions and stormwater pollutant reduction goals are described in the PCB and Mercury TMDL control plans and the MRP. Given the episodic nature of transport and the highly elusive nature of stormwater pollutant sources, management is challenging. Legacy pollutants such as PCBs are generally found throughout the urban environment because they were widely distributed across the urban landscape over many decades of use.

Bay Area municipalities and flood control districts continue to focus on balancing stormwater management with other competing public agency priorities. With regard to Bay water quality, there are three primary stormwater management and monitoring focus areas for the near future.

- Continued and enhanced implementation of stormwater control measures that significantly reduce the amount of PCBs, mercury, and other pollutants reaching the Bay. These actions include the identification and abatement of known PCB sources in the Bay's watershed, ongoing integration of green stormwater infrastructure (GSI) into the urban landscape, and management of PCB-containing materials during building demolition.

- Developing and implementing new regional water quality monitoring and modeling strategies to detect stormwater pollutant loading trends and evaluate reductions associated with control measures.

- Enhancing our understanding of the loadings of different types of CECs from stormwater and their potential impacts, and identifying stormwater control measures to address high priority CECs.

Additionally, in the decades ahead, climate change and sea level rise may imperil low-lying infrastructure and expose contaminated shoreline areas to the forces of tides and waves. This danger also presents an opportunity to renew stormwater infrastructure to prevent flooding, as well as better manage stormwater pollutants through a variety of treatment and control measures. GSI can play an important role as redevelopment occurs in the Bay Area, although it is not a panacea because redevelopment occurs slowly and does not always happen in the most contaminated areas, and because GSI is not designed to handle the most intense storms (and climate change is expected to include more extreme events). Future stormwater regulatory strategies will need to be a multifaceted combination of control measures such as consumer product bans or reformulations, cleanup of contaminated areas, and upgrading aging stormwater infrastructure to provide better management and treatment.

Integrating Green Stormwater Infrastructure into Urban Watersheds

Much of the storm drainage infrastructure in the Bay Area was constructed over 50 years ago and is currently in need of repair or replacement. The storm drains, pipes, ditches, and channels, commonly referred to as "gray" infrastructure, were designed to convey stormwater runoff away from urban areas to local creeks, channels, and the Bay as quickly as possible to avoid flooding. While gray stormwater infrastructure forms a valuable and needed foundation for effective stormwater conveyance, it provides little opportunity for the removal of pollutants before they reach the Bay.

GSI is a cost-effective, resilient approach to managing stormwater that provides many community benefits. GSI uses vegetation, soils, and natural processes to filter pollutants from stormwater. GSI includes features such as rain gardens, permeable pavements, and green streets that are widely distributed geographically and integrated into the urban landscape. GSI complements gray infrastructure and reduces and treats stormwater runoff closer to pollutant sources, while delivering environmental, social, and economic benefits.

Implementation of GSI in the Bay Area began in the early 2000s when municipalities started requiring it for development and redevelopment projects. Stormwater treatment, including GSI, is now required for moderate and large multi-family residential, commercial, and industrial development and redevelopment projects. To date, thousands of private properties throughout the Bay Area have integrated GSI into their landscapes to reduce, slow, and clean stormwater runoff. These actions will continue to assist Bay Area cities and counties in making progress towards mandated PCBs and mercury TMDL load reduction goals, while providing additional environmental and social benefits.

Given the success of GSI implementation on private properties, Bay Area cities and counties are now developing longer-term plans with an expanded geographical scope. Municipalities have recently adopted new GSI plans, which lay out strategies, targets, and tasks needed to transition traditional gray infrastructure to include GSI over the long term, and to implement and institutionalize the concepts of GSI into standard municipal engineering, construction, and maintenance practices. These plans serve as an implementation guide for cities and counties to address the water quality impacts of stormwater runoff from both private properties and public streets and sidewalks. GSI plans also include analyses that provide assurance that specific pollutant reductions (e.g., PCBs and mercury) from stormwater discharges to local creeks and San Francisco Bay will be achieved over time.

A major barrier to implementing GSI plans, however, is a lack of adequate funding. Most Bay Area municipalities have limited or no dedicated funding for stormwater management, including for the GSI implementation needed to achieve water quality goals for the Bay.

Bay Area municipalities will need billions of dollars over the next several decades to meet these goals and rebuild aging stormwater infrastructure. Funding the redesign of our stormwater drainage infrastructure will require public education and outreach to communicate the benefits to the health of our neighborhoods and water bodies that will result from these investments.

Trends in Legacy Stormwater Contaminants

Over the past two decades, considerable monitoring and modeling have been conducted by Bay Area municipalities, the Water Board, and the RMP to establish baseline stormwater loading estimates for PCBs and mercury. These estimates provide the foundation for measuring trends in the levels of pollutants transported by stormwater and assessing progress towards TMDL pollutant reduction goals. Watershed pollutant monitoring and modeling are coordinated through the STLS Team.

In 2018, the STLS Team took a significant step forward by developing the first version of the RMP's Small Tributaries Loading Trends Strategy. The Trends Strategy provides a framework and workplan for the collection of pollutant concentration and loading information to support adaptive management decisions and detect pollutant trends in small tributaries. The Trends Strategy will begin implementation in late 2019 and help link watershed management efforts, such as GSI implementation, to changes in stormwater quality. The Trends Strategy focuses on pollutants that are currently a priority (i.e., PCBs and mercury), accepting that priorities can change and that trends in other pollutants of concern (e.g., pesticides and CECs) may become important in the future. The Trends Strategy includes the compilation of information necessary to assess trends in both management actions and pollutant concentrations or loads. Current and projected trends in stormwater pollutants will be compared to the extent and magnitude of control measure implementation to support management.

Rain garden, El Cerrito (Shira Bezalel, SFEI) ▶

Investigating Emerging Contaminants in Stormwater

Recent RMP monitoring suggests stormwater is a CEC transport pathway to the Bay. In response to preliminary findings, the RMP launched a three-year special study in 2018 to evaluate the concentrations of key CECs in stormwater. Sampling sites were selected based on factors including the extent of upstream urban land uses, with an emphasis on proximity to roadways and unique land uses associated with potential contaminant sources.

Sampling will continue through 2021. Four different academic laboratories are conducting targeted analyses of CECs. In the targeted analyses, several classes of compounds will be monitored, including PFAS, ethoxylated surfactants, organophosphate esters, bisphenols, and several chemicals associated specifically with urban stormwater, such as urban use pesticides and ingredients in vehicle tires. Data analysis, interpretation, and reporting will occur in 2022.

Based on the findings of the RMP stormwater-specific CEC studies, including the recent findings of the microplastics study, Bay Area cities, counties, and flood control districts plan to continue to work with Water Board staff on identifying high priority pollutants for stormwater management. Should CECs be identified in stormwater at levels of concern, further monitoring and source evaluations are likely to be required in future stormwater NPDES permits to inform any needed management actions. §

DREDGING AND DREDGED SEDIMENT DISPOSAL

Melissa Foley
San Francisco Estuary Institute

Beth Christian
*San Francisco Bay
Regional Water Quality Control Board*

Brenda Goeden
*San Francisco Bay Conservation and
Development Commission*

Brian Ross and **Jennifer Siu**
US Environmental Protection Agency

Josh Gravenmier
Arcadis

HIGHLIGHTS

▶ Maintaining navigation channels in San Francisco Bay is vital for the Bay Area economy

▶ Disposing dredged sediment in the ocean and at upland sites removes contaminants from the Bay

▶ Potential water quality impacts at dredging and disposal sites are carefully considered and regulated

▶ Sediment, including dredged sediment, is a precious resource necessary to restore wetlands and pursue nature-based solutions to sea level rise

◀ **Dredger near Alameda Creek** (Alamy)

Dredging 101

San Francisco Bay is a critical maritime area on the west coast of the US that supports a wide variety of uses, including international trade, commercial and recreational fishing, and cruise ship, ferry, and excursion services. While the water depth at the Golden Gate Bridge is over 300 feet deep, most of the Bay is less than 18 feet deep. Dredging of the Bay began in the 1800s to create navigation channels and ports to facilitate the expansion of maritime and recreational boating activities in the Bay. An average of over three million cubic yards (MCY) of sediment has been dredged from the Bay every year since 1990. That is roughly equivalent to 200,000 dump trucks full of sediment, which parked end-to-end would encircle the Bay nearly three times! Most dredged sediment was disposed of throughout the Bay until the 1970s when disposal was limited to a small number of in-Bay disposal sites (**Figure 1**).

Dredging and disposal of dredged sediment constitute multiple routes of exposure to pollution in the Bay (**page 58**). Unlike other pollution pathways to the Bay, dredging doesn't introduce new contaminants to the Bay. The physical process of dredging sediment can uncover and release sediment-bound contaminants that are already in the Bay, such as PCBs, mercury, pesticides, and other heavy metals. In addition, the amount of sediment in the water column often increases during dredging operations, potentially exposing multiple organisms to contaminants. Disposing of dredged sediment can also result in the transport of contaminated sediment from heavily polluted sites to cleaner areas of the Bay. On the other hand, disposing of dredged sediment can also remove pollutants from the Bay, namely by disposing of sediment in the deep ocean or containing it at upland sites.

The majority of the dredging in the San Francisco Bay area is done for operations and maintenance of federal navigation channels by the US Army Corps of Engineers (USACE). There are six deep-draft projects (**Figure 1**; Oakland Harbor, Redwood City Harbor, Richmond Inner and Outer Harbors, San Francisco Harbor, San Pablo Bay and Mare Island Strait, and Suisun Bay Channel) and five shallow-draft projects (Jack P. Maltester Channel [San Leandro Marina], Napa River, Petaluma River, San Rafael Creek, and Suisun Slough Channel). Many additional non-federal dredging projects, including refinery wharfs and small marinas, constitute about a third of dredged volumes annually.

Maintaining navigational channels and harbors is critically important to the Bay Area economy and allows ships to safely navigate into and out of ports, harbors, and marinas without running aground. Over 400 vessels move around the San Francisco Bay every day. Combined vessel imports and exports in San Francisco Bay exceeded $68 billion in 2017. The cargo tonnage at the Port of San Francisco and the number of cruise ships entering the Bay have increased by 30% in the last five years, while approximately 2.5 million containers move through the Port of Oakland annually, the third largest volume on the west coast.

Regulatory Framework

The Long-Term Management Strategy (LTMS) for the Placement of Dredged Material in the Bay Region was formed in 1990 after controversies and environmental impacts highlighted the need for improved coordination and management of dredging activities. The LTMS is made up of federal and state agencies that regulate dredging and disposal activities in San Francisco Bay, along with representatives from the dredging, environmental, regulatory, and scientific communities. The LTMS management plan was published in 2001, outlining policies and measures for implementing the LTMS (LTMS 2001). The 2008 Pulse of the Estuary provides a more in-depth discussion of the formation and structure of the LTMS (Delaney et al. 2008).

The LTMS has accomplished a lot since its inception. The 12-year review of the program in 2013 re-confirmed the LTMS goals while calling for

**San Francisco Bay
Operations and Maintenance
Dredging Projects and Disposal Sites**

PETALUMA RIVER

NAPA RIVER

SUISUN SLOUGH
CHANNEL

SF-9

SF-16

SF-10

SAN PABLO BAY &
MARE ISLAND STRAIT

SUISUN BAY CHANNEL

SAN RAFAEL CREEK

RICHMOND
HARBOR

LARKSPUR FERRY
CHANNEL

OAKLAND
HARBOR

SF-11

SAN FRANCISCO
HARBOR

SF-8

SF-17

JACK D.
MALTESTER
CHANNEL
(SAN LEANDRO
MARINA)

REDWOOD
CITY HARBOR

**Figure 1. An average of over three
million cubic yards of sediment
has been dredged from the Bay
every year since 1990** to maintain
and expand navigational channels
(red). The dredged material is
transported and released at
several sites in the Bay and ocean
(green), or beneficially reused in
restoring wetland and other habitat,
stabilizing levees, or capping and
lining landfills.

APPROX. 50 MILES
FROM GOLDEN GATE

SF-DODS

5 miles

5 km

N

Dredging: Sources, Pathways, Loading

Dredging removes bottom sediment from navigational channels in the Bay and transports it to disposal sites in the Bay, the ocean, upland, or to habitat restoration projects for "beneficial reuse". The **SOURCES** of pollutants in dredged material are all of the sources that lead to contamination of Bay sediment. Stormwater sources are often dominant contributors to dredged material pollution, due to the proximity of dredging sites to stormwater inputs.

Disposing of dredged sediment at in-Bay disposal sites results in a relocation within the Bay, not a net loading. Disposal of dredged sediment in the ocean or at upland sites, or re-using it in wetland restoration projects, removes pollutants from circulation in the Bay – for example, 50% of the PCB mass that is dredged is removed.

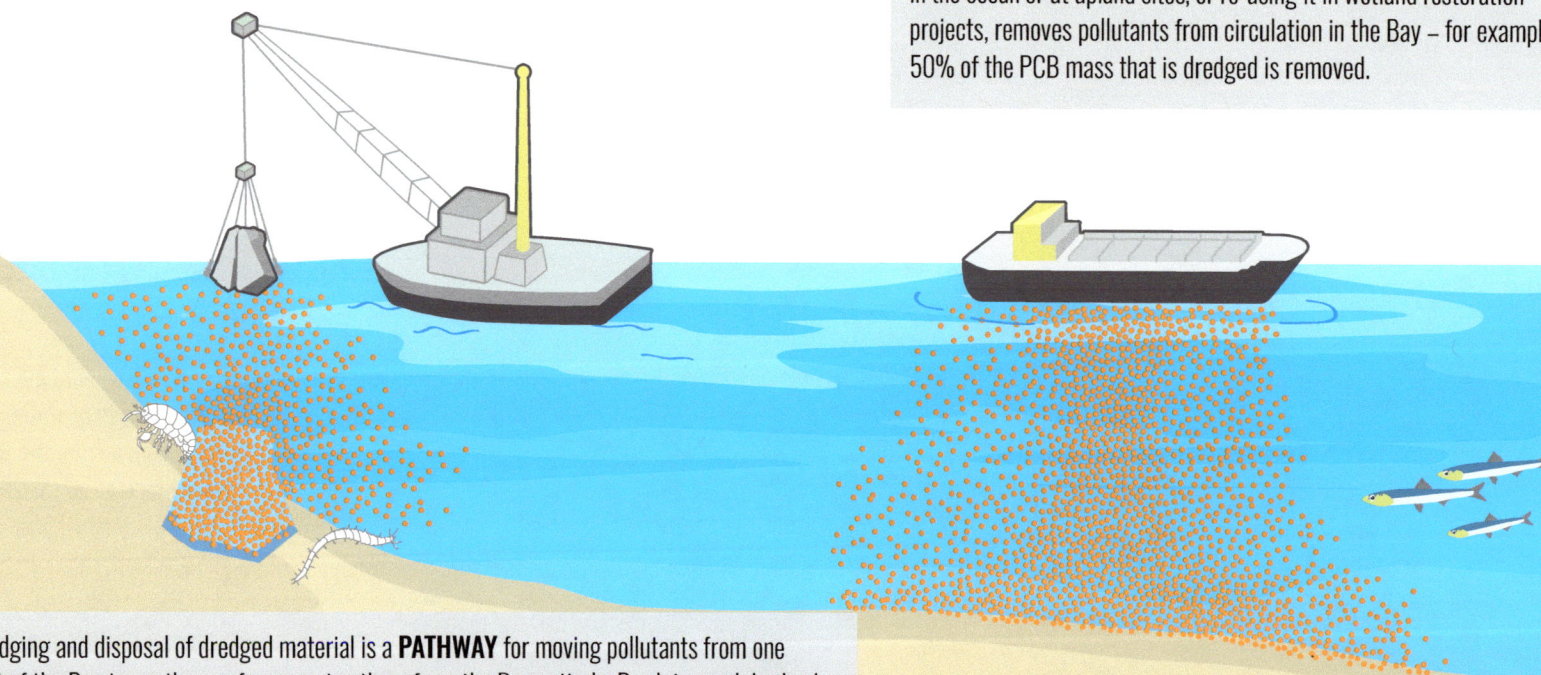

Dredging and disposal of dredged material is a **PATHWAY** for moving pollutants from one part of the Bay to another, or for removing them from the Bay entirely. Dredging and dredged material disposal uncover and remobilize sediment-bound contaminants, such as PCBs and mercury. In addition, the amount of sediment in the water column increases during dredging and disposal operations, potentially increasing the exposure of aquatic organisms to pollutants.

Illustration by Linda Wanczyk (lindawanczyk.com)

more flexibility in meeting them over time (LTMS 2013). One of the primary goals of the LTMS was to significantly reduce in-Bay disposal by 2012. During the transition period from 2001 to 2012, in-Bay disposal volumes were successfully reduced from 2.8 MCY to the target of 1.25 MCY. Since 2012, the amount of in-Bay disposal has continued to fall below the 1.25 MCY limit set forth in the LTMS (**Figure 2**).

There are four in-Bay disposal sites (**Figure 1**), including one each in Suisun Bay (SF-16), Carquinez Strait (SF-9), San Pablo Bay (SF-10), and near Alcatraz Island (SF-11). The Alcatraz disposal site tends to be the most frequently used in-Bay disposal location, followed by San Pablo Bay (SF-10). Ocean disposal and beneficial reuse totals are highly variable from year to year. Beneficial reuse totals ranged between 0.5 and 1.2

MCY per year from 2012 to 2017; ocean disposal totals spanned an order of magnitude during that same period from 0.1 to 1.6 MCY. The amount of dredged sediment disposal at the different locations is highly dependent on the total amount of sediment dredged each year and the location of that dredging.

In 2011, the LTMS also initiated a programmatic consultation on Essential Fish Habitat and an amendment to the programmatic Endangered Species Act consultation with the National Marine Fisheries Service (NMFS), a division of the National Oceanic and Atmospheric Administration (NOAA). The Essential Fish Habitat consultation removed the need for individual consultations for most dredging projects and improved the predictability of sediment testing by establishing

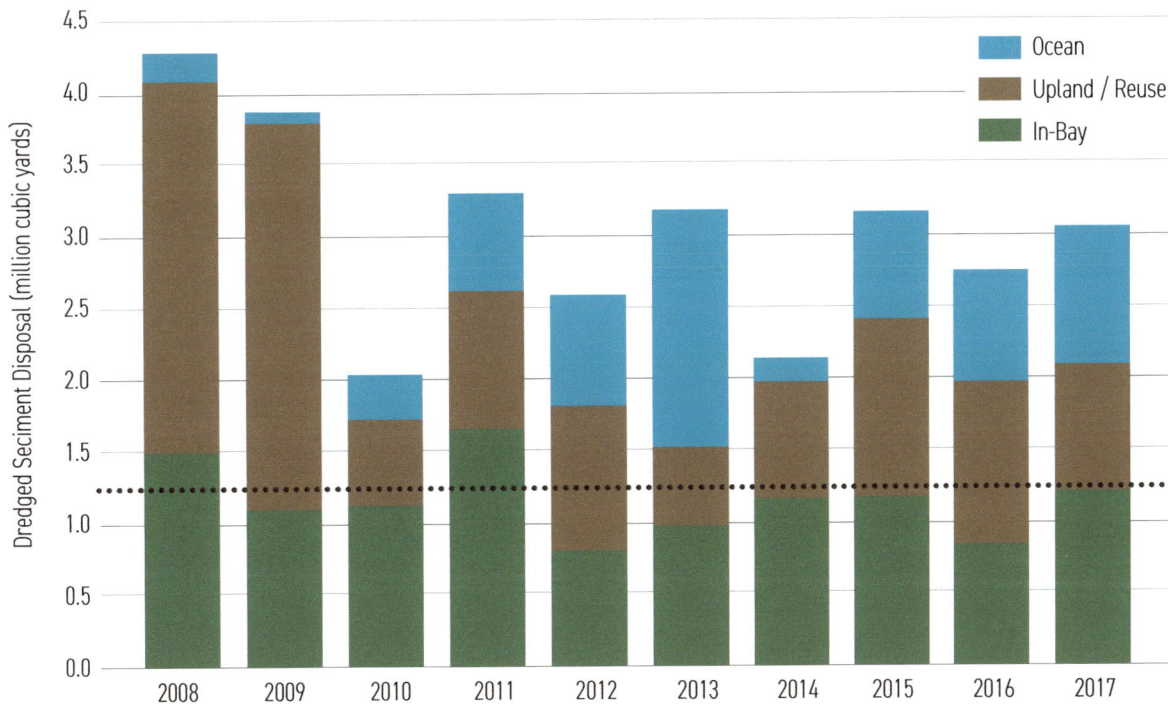

Figure 2. One of the primary long-term goals for management of dredged material in the Bay was to significantly reduce in-Bay disposal by 2012. During a transition period from 2001 to 2012, in-Bay disposal volumes were successfully reduced from 2.8 MCY to the target of 1.25 MCY. Since 2012, the amount of in-Bay disposal has continued to be below the 1.25 MCY limit set forth in the LTMS.

FOOTNOTE: Dredged sediment disposal volume by location and year. The dashed line represents the LTMS target of 1.25 MCY for in-Bay disposal.

bioaccumulation testing trigger thresholds. In addition, it significantly improved eelgrass habitat protection measures during dredging operations. The amendment to the programmatic Endangered Species Act consultation also resulted in refinements to dredging restrictions, as well as support for additional beneficial reuse, while allowing more flexibility to work outside environmental work windows. Projects that propose to dredge outside the salmon work window can do so in many cases, as long as dredged sediment is taken to a site that will benefit fish habitat.

The Dredged Material Management Office (DMMO), another element of the LTMS, is a joint program of federal and state agencies that coordinate the review of sediment quality sampling plans, analyze the results of sediment quality sampling, and make disposal and placement suitability determinations for proposed dredging projects in the Bay. Suitability is based on physical characteristics (grain size, total organic carbon, and total solids), bulk sediment chemistry (10 metals, 4 butyltins, 25 PAH compounds, 22 organochlorine pesticides, 40 PCB congeners, and dioxins), and benthic and water column toxicity testing. Suitability is also measured against bioaccumulation triggers and

total maximum daily load (TMDL) values for seven contaminant classes: mercury, total PCBs, total PAHs, total DDTs, total chlordane, dieldrin, and dioxins. Thresholds for mercury, PCBs, and PAHs are based on Bay ambient sediment contaminant concentrations determined via the Regional Monitoring Program for Water Quality in San Francisco Bay (RMP) and are updated as new data become available. The sediment testing program also meets the national testing guidelines of the Inland Testing Manual (USEPA and USACE 1998) and the Ocean Testing Manual (USEPA and USACE 1991), and allows less frequent testing where existing data consistently show contaminant levels are below the screening guideline values.

The DMMO built a database to house and make sediment testing data publicly available (www.dmmosfbay.org). The database currently includes data from 2000 to 2018 and allows new project data to be directly uploaded to the site. This database has improved the efficiency of permit review, and has allowed for additional data analysis. Funds from the RMP are being used to maintain the database and improve querying capabilities.

Container ships at the Port of Oakland ▶
(Shira Bezalel, SFEI)

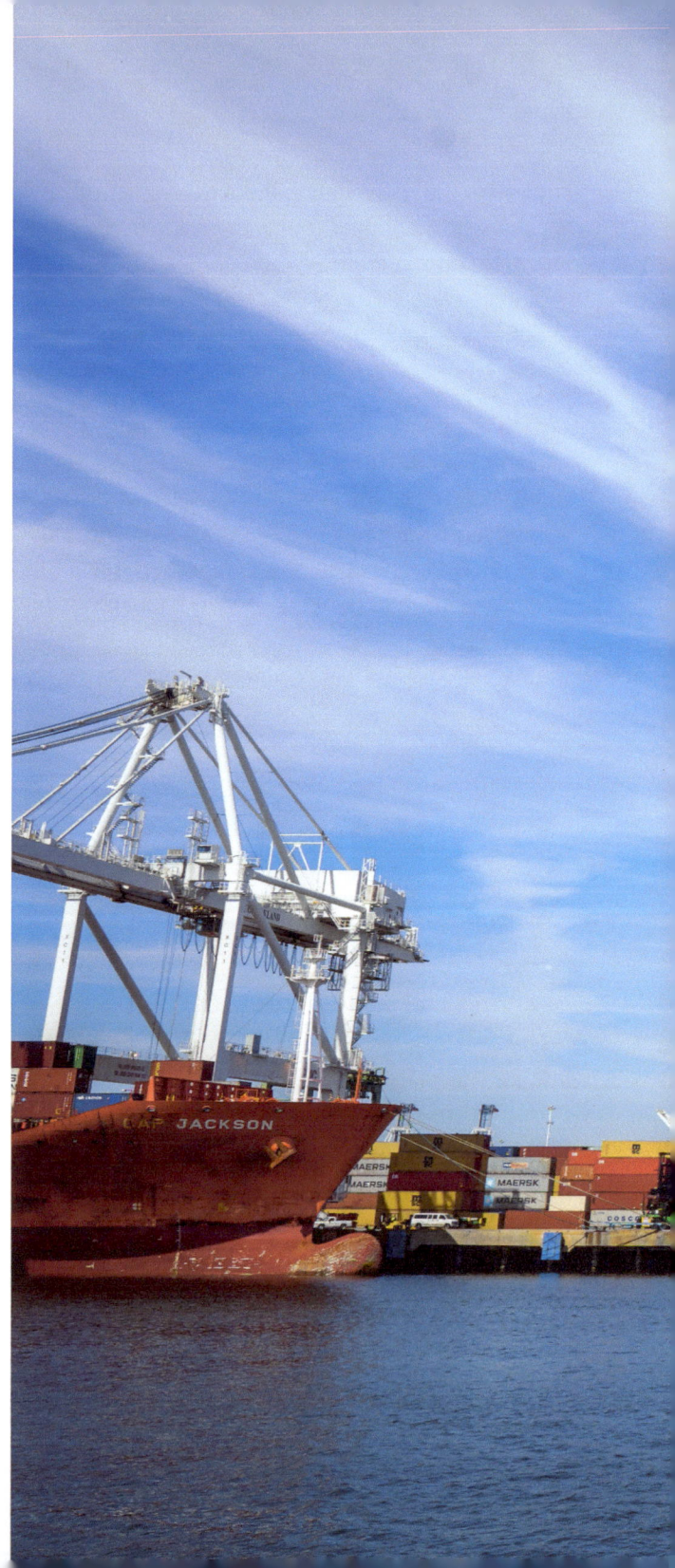

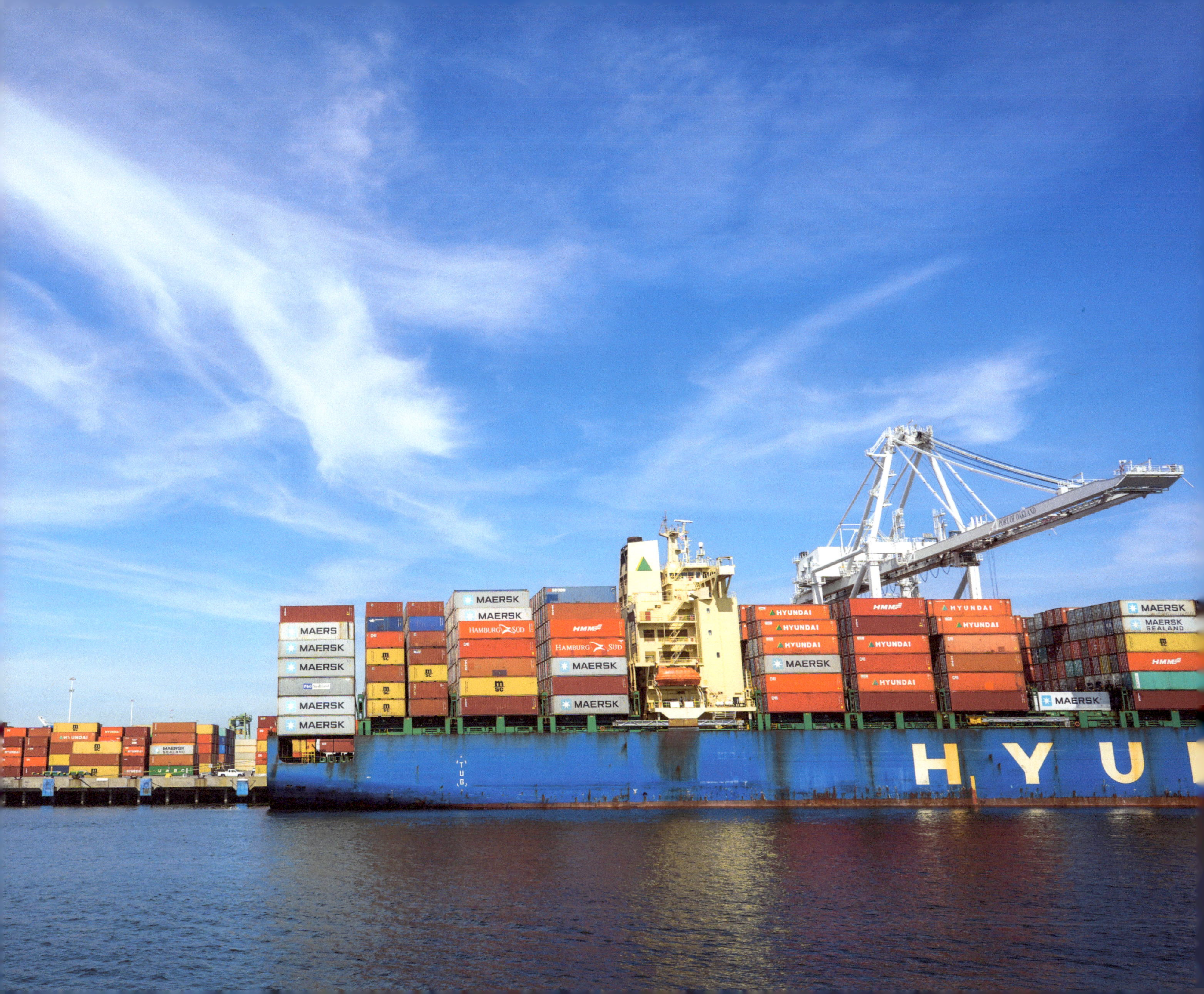
MAERSK
HAMBURG SÜD
HMM
HYUNDAI
MAERSK SEALAND

Recent Studies

There have been a number of recent studies conducted by the RMP and partners supporting the management of dredging. These studies range from understanding the impacts of dredging, to updating ambient sediment contamination levels, to evaluation of bioaccumulation thresholds.

USGS Study of Dredging Site Impacts

Using funding from the RMP, Port of San Francisco mitigation, and the USACE, the USGS assessed the impact of dredging on macroinvertebrate populations at marinas in Central San Francisco Bay. Macroinvertebrate density was 14 to 167% greater in undredged reference sites compared to dredged areas at four of five study sites two to three years following dredging activity (De La Cruz et al. 2018). The types of taxa found at dredged sites differed from those found at undredged sites, suggesting that changes in sediment characteristics and/or disturbance likely affected macroinvertebrate community composition.

Bay Ambient Sediment Contaminant Concentrations

Where available and appropriate, Bay-specific ambient sediment contaminant concentrations (i.e., concentrations from less polluted areas of the Bay) are used to assess dredged sediment for disposal options. Ambient sediment concentrations are determined using RMP sediment samples and calculating the 90th percentile value (excluding the worst 10% of data) for the Bay. These concentration values are updated and recalculated as new data are available. Ambient values for contaminants subject to TMDL limits or bioaccumulation testing thresholds, including mercury, total PCBs, and total PAHs, have been updated several times since 2011 but have remained relatively constant. In 2015, the RMP updated the ambient sediment contaminant concentrations for an additional 91 contaminants routinely monitored by the RMP. Previous concentrations were based on a smaller subset of data from 1991 to 1995. For a majority of contaminants, data from randomized sites from 2003 to 2012 were used to calculate ambient concentrations (Yee et al. 2015). Ambient concentrations provide a benchmark by which to compare dredged sediment, but they are not necessarily indicative of environmentally desirable conditions.

Sediment Bioaccumulation Trigger Updates

The USEPA Region IX and the USACE reviewed sediment mercury and bioaccumulation data to look for a relationship between the two factors (Ross 2012). There was no relationship between increasing mercury concentration in the sediment and bioaccumulation in exposed organisms. As a result of this analysis, the bioaccumulation trigger was eliminated for mercury, thereby reducing testing costs for dredgers. The RMP will be evaluating the relationship between PCBs and sediment bioaccumulation data in 2020.

DMMO PCB Synthesis

The PCB concentrations at dredged sites contribute modestly to overall PCBs in the Bay. Between 2000 and 2017, PCB concentrations in dredged sediment were higher than in open Bay sediment but were similar to sediment in the Bay margins (Yee et al. 2019a). The estimated PCB mass moved out of the Bay (deep ocean, upland disposal, or beneficial reuse) via dredged sediment disposal was just over 50% of the PCB mass encountered in dredged sediment, confirming a key assumption of the PCB TMDL.

DMMO Dioxin Information

Dioxin concentrations in sediment from ports, marinas, and other nearshore areas less than 250 m from the shoreline were higher than in sediment from open Bay areas (Yee et al. 2019b). These areas contribute a small proportion of the overall dioxin load in the Bay, but represent a disproportionate source based on the size of the dredged areas. Still, only limited areas have dioxins present at concentrations that could limit disposal or reuse of the dredged sediment.

Toxicity Reference Values

Toxicity reference values (TRVs) are used as a screening tool to evaluate whether the concentrations of contaminants in benthic organisms observed in bioaccumulation assays could have adverse ecological effects. TRVs for PCBs and DDTs were identified by the RMP for the San Francisco Bay using the current DMMO methodology for selecting TRVs (Lin and Davis 2018). TRVs for other compound classes could not be identified due, in part, to the DMMO approach for screening potential TRVs. A follow-up study, using an expanded screening methodology, has been proposed.

Sediment Supply to San Francisco Bay

A synthesis of the best available data for sediment was completed in 2018 (Schoellhamer et al. 2018) to inform ongoing sediment monitoring efforts in the Bay. The report also provided the first regional estimate of sediment supply to the Bay for an extended period (1996-2016). An estimated 1.9 million metric tonnes per year of sediment entered the Bay during that time, with over 60% being supplied by small tributaries. The remaining sediment was supplied through the Delta, as measured at Mallard Island by RMP-funded USGS monitoring. The source of sediment affects where it is deposited in the Bay. Sediment from small tributaries is likely to be deposited in tidal channels and Bay margins rather than the open Bay.

Loading cranes and container ships ▶
at the Port of Oakland (Shira Bezalel, SFEI)

Beneficial Reuse Projects

The LTMS outlines the long-term goal of using at least 40% of dredged sediment for beneficial purposes, which includes restoring marshes and wetlands, creating in-Bay habitat, stabilizing levees, and capping and lining landfills. Based on the San Francisco Regional Water Quality Control Board's sediment screening and testing guidelines for dredged sediment (SFBRWQCB 2000) and US Fish and Wildlife biological opinions for some restoration sites, non-toxic dredged sediment that falls below sediment chemistry screening guidelines can be beneficially used for wetland restoration. To date, over 25 million cubic yards of dredged sediment have been used to restore over 7,500 acres of wetlands, including Sonoma Baylands, Hamilton Wetlands, Bair Island, Cullinan Ranch, and Montezuma Wetlands (**Figure 3**). Cullinan Ranch and Montezuma Wetlands, two large projects in San Pablo Bay, received over 8 MCY of beneficial reuse sediment between 2003 and 2017. The Middle Harbor Enhancement Project at the Port of Oakland used dredged sediment from the harbor deepening project to restore nearby subtidal habitat at the Port. These projects have resulted in 25 to 60% of dredged sediment per year being used for beneficial reuse over the last ten years.

Providing dredged sediment to beneficial reuse projects is often a more expensive option than aquatic disposal, due in part to the distance between dredging locations and restoration projects. Additional equipment and staffing are also required to offload dredged sediment from barges onto the restoration site. There are new funding opportunities that have recently come online that may help alleviate some of these costs for dredgers. The Bay Restoration Authority is now distributing funds from Measure AA that were authorized in 2016 by voters in the nine-county Bay Area. Over the next 20 years, approximately $500 million will be collected via a parcel tax to support Bay water quality improvement, pollution prevention, and habitat restoration projects. Strategic outcomes identified for Measure AA funds include restoring 15,000 acres of wetland and tidal marsh and constructing 20 miles of levees to facilitate wetland restoration projects. The LTMS agencies are currently working with the Restoration Authority to identify ways to support these efforts and fund beneficial reuse of dredged sediment.

A second source of new funding for beneficial reuse of dredged sediment comes from the State of California, through the work of San Francisco Baykeeper, California Coastkeeper Alliance, and the Bay Conservation and Development Commission (BCDC) with Assembly Member Kevin Mullin. For the first time, the State is providing funding for the incremental cost of placing dredged sediment at a wetland restoration site. Beginning in summer 2019, the Coastal Conservancy will spend $6 million to implement a five-year project to move dredged sediment from Redwood City Harbor to nearby marsh restoration sites.

An additional effort to secure funding for beneficial reuse of dredged sediment was recently proposed by the Coastal Conservancy to the USACE. In 2016, the Coastal Conservancy proposed a project for funding under the Water Resources Development Act 2016 (Water Infrastructure Improvements for the Nation) Section 1122 USACE Beneficial Reuse of Dredge Material Pilot Program to use dredged sediment from four federal navigation channels to restore and create habitat at four tidal wetland restoration sites. Approximately 1.5 to 2.5 MCY of sediment is dredged annually from the Oakland, Richmond, and Redwood City harbors, as well as the Pinole Shoal Channel. USACE approved a portion of the proposal to test a sediment placement method known as strategic placement, but did not choose to fund the habitat restoration component of the proposal. Using this method, sediment is placed in the Bay nearshore environment with the expectation that tides and currents will wash it into wetlands to augment sediment supply as sea level rises. While this novel approach needs to be tested and is an important piece of the larger wetland restoration strategy, the region has also recognized the critical need to place large quantities of sediment directly at restoration sites to reach elevations necessary for marsh development. The Coastal Conservancy and BCDC are pursuing opportunities to secure additional money for USACE to implement the full proposal.

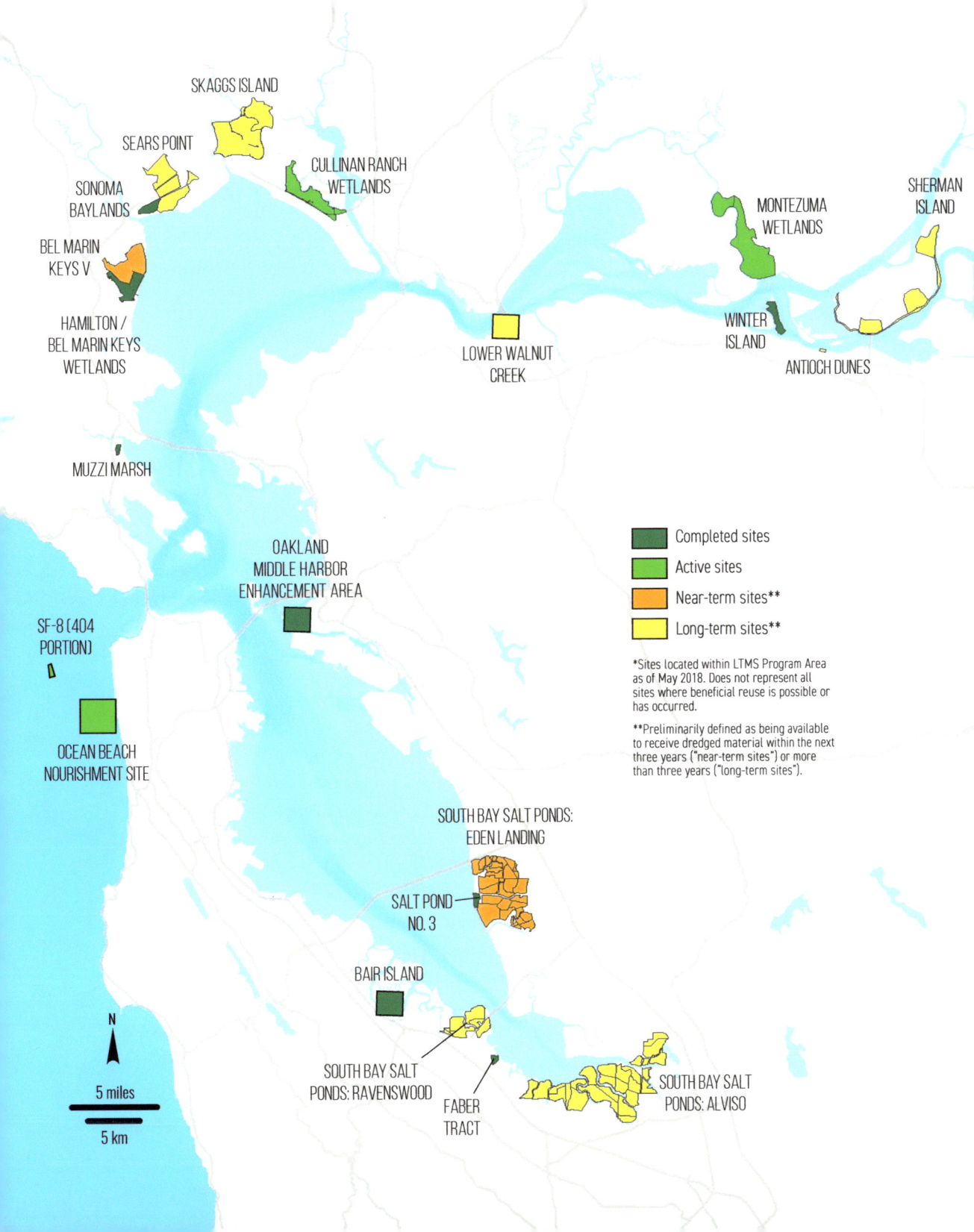

SKAGGS ISLAND

SEARS POINT

SONOMA
BAYLANDS

BEL MARIN
KEYS V

HAMILTON /
BEL MARIN KEYS
WETLANDS

CULLINAN RANCH
WETLANDS

MONTEZUMA
WETLANDS

SHERMAN
ISLAND

WINTER
ISLAND

ANTIOCH DUNES

LOWER WALNUT
CREEK

MUZZI MARSH

OAKLAND
MIDDLE HARBOR
ENHANCEMENT AREA

SF-8 (404
PORTION)

OCEAN BEACH
NOURISHMENT SITE

Completed sites

Active sites

Near-term sites**

Long-term sites**

*Sites located within LTMS Program Area
as of May 2018. Does not represent all
sites where beneficial reuse is possible or
has occurred.

**Preliminarily defined as being available
to receive dredged material within the next
three years ("near-term sites") or more
than three years ("long-term sites").

SOUTH BAY SALT PONDS:
EDEN LANDING

SALT POND
NO. 3

BAIR ISLAND

SOUTH BAY SALT
PONDS: RAVENSWOOD

FABER
TRACT

SOUTH BAY SALT
PONDS: ALVISO

N

5 miles

5 km

**Figure 3. Long-term Management
Strategy (LTMS) beneficial reuse
placement sites in San Francisco
Bay*.** Another major long-term goal for
management of dredged material in the Bay
is to use at least 40% of dredged sediment
for beneficial purposes, which includes
restoring marshes and wetlands, creating in-
Bay habitat, stabilizing levees, and capping
and lining landfills. To date, over 25 million
cubic yards of dredged sediment have been
used to restore over 7500 acres of wetlands.
Map adapted from LTMS (2018).

Future Challenges

Beneficial Reuse

Because San Francisco Bay is a shallow harbor surrounded by a vibrant community that relies on the maritime industry as an economic driver, maintenance dredging of federal navigation channels, ports, oil terminals, and marinas will undoubtedly continue into the future. As ships continue to grow larger, there is potential that additional deepening of these areas will be proposed. A portion of the federal navigation channel leading to the Ports of Stockton and Sacramento is currently proposed for deepening. Deepening projects produce a large quantity of sediment initially and require additional maintenance dredging into the future. Whether changes brought by climate change, and specifically rising seas, will affect dredging efforts remain to be seen. Concerns over issues such as salinity intrusion into the Delta (an important source of fresh water for California) and changes in biological communities will factor into these decisions. It is possible that rising seas may reduce the need for maintenance dredging over time, but this has not yet been considered in regional planning. What remains true is that the region has a great need for sediment to restore the tidal wetlands that were lost to development and bolster the remaining wetlands that are threatened by rising seas.

Beneficial reuse of dredged sediment at wetland restoration sites is a key ingredient to the Bay Area's sea level rise resilience strategy. As the Baylands Ecological Habitat Goals Update (Goals Project 2015) highlighted, reusing dredged sediment is a key component in meeting habitat restoration and endangered species goals, as well as building natural infrastructure to address coastal flooding. It is the only large source of sediment identified that provides the Bay's own sediment to support wetland vegetation. While the LTMS has had substantial success in reusing a significant portion of dredged sediment for tidal wetland restoration, challenges remain. The primary barriers to beneficial reuse are the availability of offloading equipment at the restoration sites, funding for operational expenses to place the dredged sediment,

and a lack of restoration sites permitted to receive sediment. If the region were to invest in one or two multi-user sediment offloaders and operation expenses, this would substantially reduce costs to individual dredgers and remove multiple barriers to sediment reuse. Restoration site managers should consider using dredged sediment to raise site elevations to establish marsh vegetation quickly. While this factor increases planning and site management costs, it is likely the only way deeply subsided sites will reach marsh elevation and begin vegetating before rising seas makes this an impossibility. A recent analysis by Perry et al. (2015) suggest that over 150 million cubic yards of sediment would be needed to bring the 40,000 acres of planned or in-progress marsh restoration areas up to current marsh plain elevation. This does not account for sea level rise, which would potentially require additional sediment on the order of millions of tonnes to maintain those restored marshes into the future. Despite this need for sediment, there are currently only two sites using dredged sediment: Cullinan Ranch and Montezuma. Eden Landing and Bel Marin Keys V are on the horizon but have not yet sought permits for construction.

An opportunity exists for the USACE to increase beneficial reuse of sediment dredged from the federal navigation channels. USACE is currently scoping its regional dredged sediment management plan to assess the dredging needs and dredged sediment disposal capacity at current sites for the next twenty years. This offers the opportunity for the USACE to incorporate beneficial reuse of dredged sediment into their operations to deliver sustainable economic, environmental, and social benefits to the Bay Area consistent with the Engineering with Nature initiative. However, USACE has funding limitations which force it to utilize the more cost effective ocean and in-Bay disposal locations. Those sites are less desirable because deep ocean disposal does not provide the benefits of reuse and removes sediment resources from the Bay. In-Bay disposal sites, while keeping sediment in the system, are not located close enough to nearshore areas were sediment is needed most. The USACE Regional Dredge Material Management Plan creates an opportunity to identify and encourage future beneficial reuse disposal sites and new partners to meet the needs of the region while maintaining navigation safety for the maritime industry.

Information Needs

Connecting dredging operations to beneficial reuse projects is an important component of increasing the use of dredged sediment for beneficial reuse. SediMatch is an online tool that provides a framework and information source for match-making between restoration projects and navigational and flood protection dredging projects (sedimatch.sfei.org). SediMatch is a collaborative program of the San Francisco Bay Joint Venture, San Francisco Estuary Partnership, BCDC, and SFEI. Data on sediment need or availability collected through SediMatch has the potential to contribute to the understanding of sediment distribution and movement throughout the estuary system, helping scientists better target their research and assisting managers with decision-making related to current and future conditions.

As sea level continues to rise, marshes and tidal wetlands may not be able to keep pace with sea level rise. Using dredged sediment to give these habitats a boost is critically important. Revisiting the current sediment chemistry guidelines to ensure they are appropriate and not unnecessarily conservative will be an important component of ongoing work by the LTMS and RMP. §

Clamshell dredge and barge (Alamy) ▶

WATER QUALITY
UPDATES

pages 68-93

◀ **View of the Golden Gate from Angel Island** (Shira Bezalel, SFEI)

GRAPH DETAILS ON PAGE 96

Delta Sediment Load

Flows from the Delta provide a large proportion of the Bay's sediment supply. Delta sediment loads are highly variable from year to year due to fluctuation in rainfall. Delta sediment loads were relatively low from 2012-2016 due to a five-year drought. Record rainfall in the watershed led to a high load in 2017 - the highest since 1998. This was followed in 2018 by one of the lowest sediment loads over the period of record.

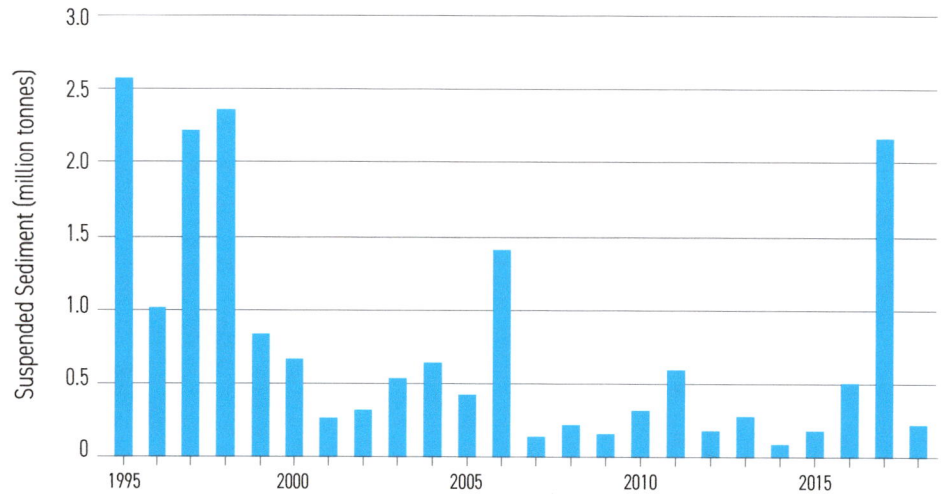

Suspended Sediment

Suspended sediment particles in the Bay are a source of sediment for wetland restoration, affect sunlight penetration and algae growth in the Bay, and are a vehicle for transport of pollutants. A sudden Bay-wide shift in suspended sediment occurred in the late 1990s, and concentrations remained low for the next 10 years, as indicated in this plot for a station at the Dumbarton Bridge, and leading to a hypothesis that the pool of erodible sediment in the Bay had been depleted. Levels were higher, however, at this location in 2014-2018, indicating that the erodible pool in Lower South Bay is still present and available for resuspension.

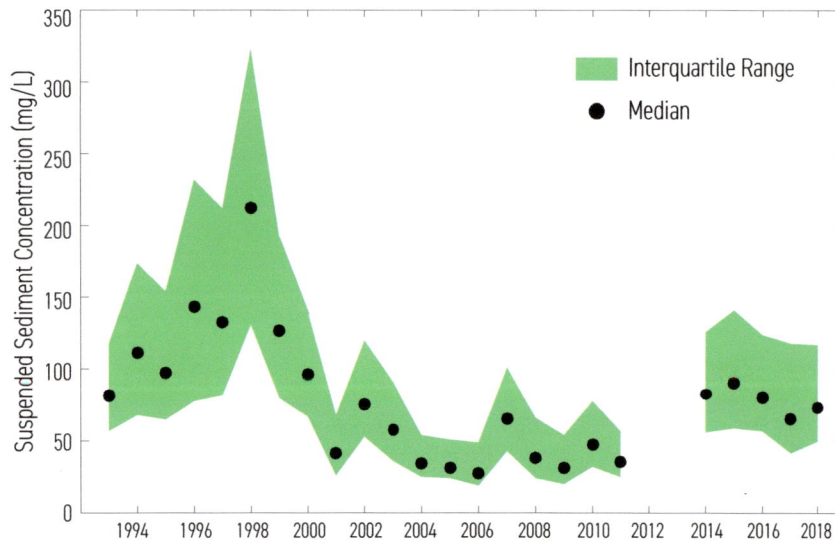

Restored Wetland

Tidal wetlands are part of the Bay. They are intimately connected to the open waters of the Bay through the exchange of water and sediment and the movement of aquatic species, and have a strong influence on Bay water quality. The ambitious plan to restore 100,000 acres by the year 2100 will add an area equivalent to one-third of the surface area of the Bay. Almost 17,000 acres have been restored since 1993.

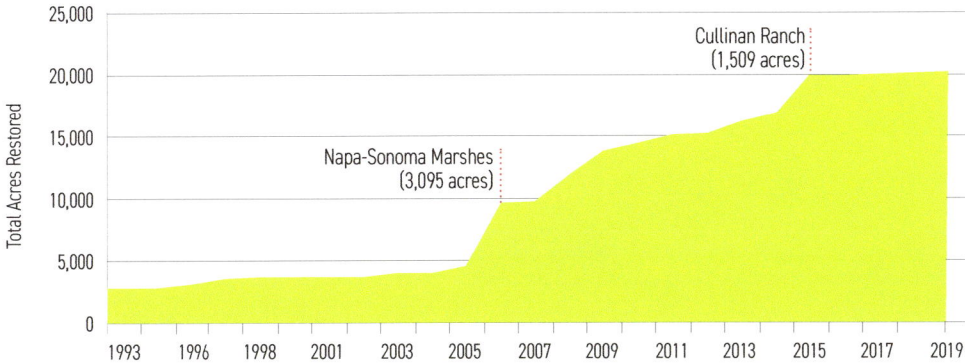

FOOTNOTE: Data summarized from Project Tracker (ptrack.ecoatlas.org).

In-Bay Disposal

In 2000, state and federal agencies adopted a Long-Term Management Strategy to reduce in-Bay disposal of dredged material and to maximize the beneficial reuse of dredged material. Beneficial reuse includes constructing wetland restoration projects, levee repair, and use as construction fill. The LTMS Plan called for reduction of aquatic disposal in the Bay to approximately 1.25 million cubic yards per year by 2012. This goal was met in every year from 2012-2017. The in-Bay disposal volume in 2017 was 1,220,000 cubic yards.

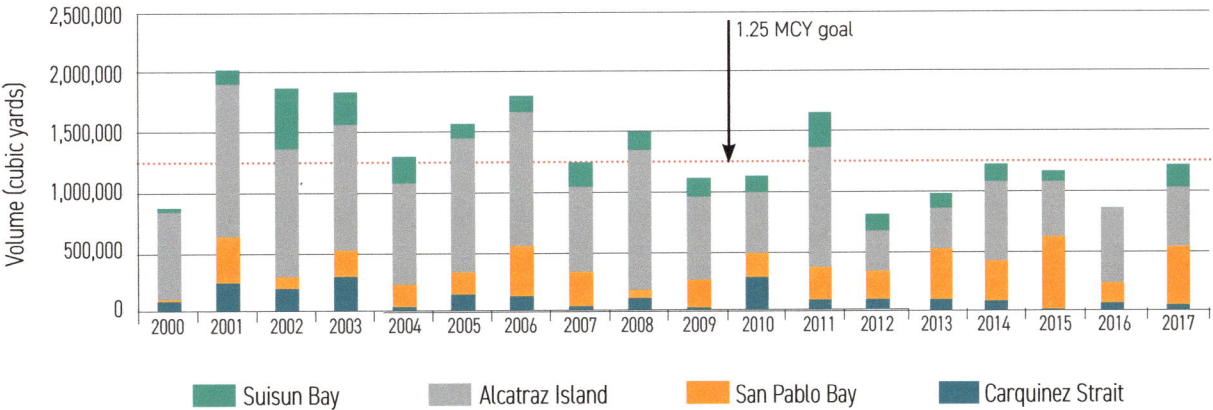

FOOTNOTE: Data source: Dredged Material Management Office annual reports and records.

GRAPH DETAILS ON PAGE 96

Sea Level

Rising sea level will affect Bay water quality in many ways, through its influence on the evolution of shoreline habitats and on pollutant fate in Bay waters. A tide gauge at the Golden Gate Bridge has been in operation since 1854, making it the nation's oldest continually operating tidal observation station and providing the longest continuous tide record in the Western Hemisphere. Based on a 20-year rolling average, sea level at the Golden Gate rose 7.1 inches (0.18 meters) from 1916-2018.

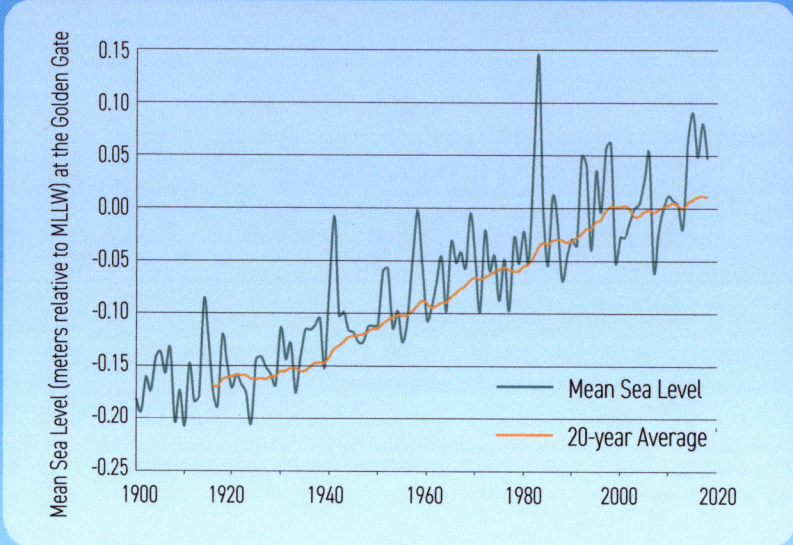

Chlorophyll

Excessive increases in phytoplankton abundance in response to elevated nutrient concentrations are common in estuaries around the world. To date, the Bay has exhibited resistance to the large algal blooms and resulting low dissolved oxygen that have plagued other nutrient-enriched estuaries. Chlorophyll concentrations in water provide an index of the abundance of phytoplankton and a key means of tracking whether a problem may be developing.

Chlorophyll concentrations in South Bay and Lower South Bay have increased since the mid-1990s. Cloern et al. (2007) first documented increasing fall chlorophyll concentrations in South Bay, with approximately a 2.5-fold increase between 1995 and 2005. The trend of increasing chlorophyll led to concerns that South Bay's resistance to nutrients was declining. At that point it was unclear whether phytoplankton biomass would continue increasing or stop. Data collected after 2005 indicate that phytoplankton biomass has stopped increasing and reached a new plateau, but at a higher level than the concentrations that prevailed from 1980 to 1995.

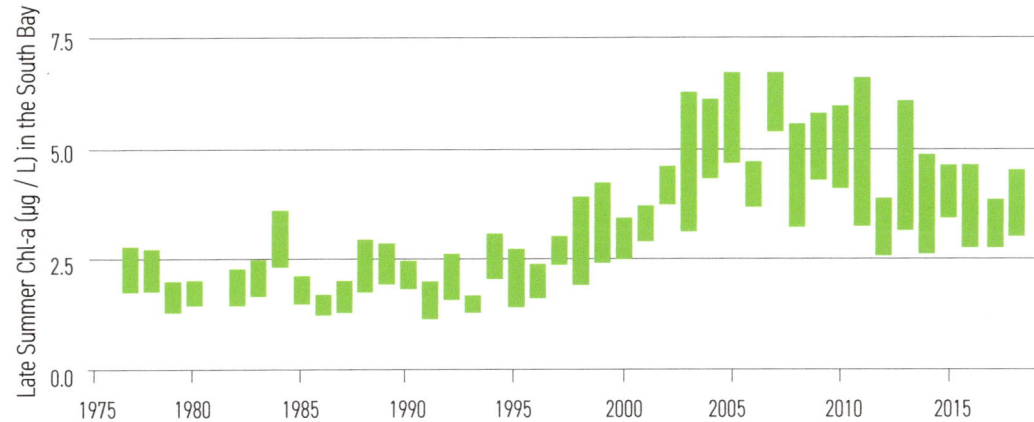

FOOTNOTE: The middle range (between the 25th and 75th percentiles) of annual chlorophyll concentrations in the South Bay in late summer.

◀ **Tide gauge station at the Golden Gate Bridge**
(Shira Bezalel, SFEI)

The R/V David Peterson ▶
(Amy Richey, SFEI)

WATER QUALITY UPDATES

73

Long-term Trends in Important Nutrient Parameters

The following graphs provide summaries of several parameters that are relevant to nutrient management in the Bay. The data illustrate the substantial variability in these parameters by season, year, and region in the Bay, and why capturing this variability requires a robust observational program.

Nitrogen (N) and phosphorus (P) are natural and essential components of healthy estuarine ecosystems. Sufficient nutrient levels are needed to support the growth of phytoplankton (microscopic floating algae) that in turn serves as the base of the food web. Too much N or P, however, can lead to unhealthy levels of phytoplankton depending on other factors such as water clarity, temperature, and vertical mixing. The concentrations of N and P are generally highest in South Bay, reflecting the large discharges of nutrients from wastewater treatment facilities and limited flushing in this portion of the Bay. N and P concentrations vary seasonally at locations throughout the Bay, with concentrations regulated by a balance between multiple processes (inputs, uptake by phytoplankton, microbial transformations, and transport).

Photic depth, a measure of water clarity, is the depth at which light levels are reduced to 1% of incident light. Higher values of photic depth indicate greater water clarity. The clearest waters in the Bay are in Central Bay, where a photic depth of 5 meters is common. In contrast, the waters of South Bay tend to be turbid with a photic depth of only 1-2 meters. The thin photic layer in South Bay is one of the factors that limits algae blooms in this area despite the high nutrient concentrations.

Chlorophyll-a is a measure of phytoplankton abundance. While South Bay has historically experienced sizable spring phytoplankton blooms (Cloern and Jassby 2012), major blooms have been notably and inexplicably absent over the past several years, except for a short-lived peak observed at South and Lower South Bay stations in February 2013. An increase in fall chlorophyll-a levels in South Bay, observed beginning in the late 1990s through 2005 (Cloern et al. 2007), was among the original motivations for the Water Board to establish the Nutrient Management Strategy. This indicator continues to be tracked (**page 73**), and observations through 2015 suggest that fall chlorophyll-a levels have leveled off.

NUTRIENTS

GRAPH DETAILS ON PAGE 96

STATION

- ■ s6
- ■ s13
- ■ s18
- ■ s21
- ■ s27
- ■ s32
- ■ s36

	NORTH BAY	CENTRAL BAY	SOUTH BAY
Dissolved Inorganic Nitrogen (µmol/L)	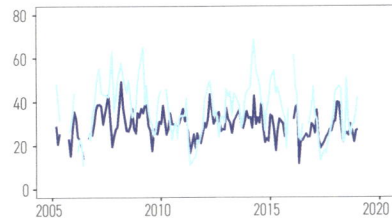	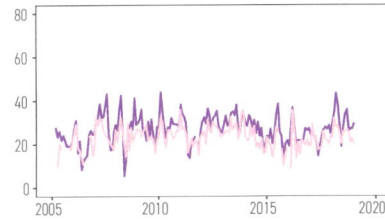	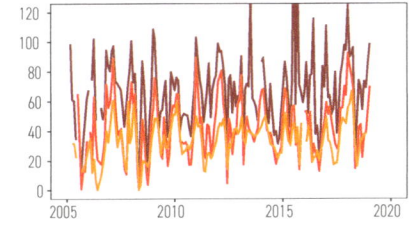
Phosphate (µmol/L)	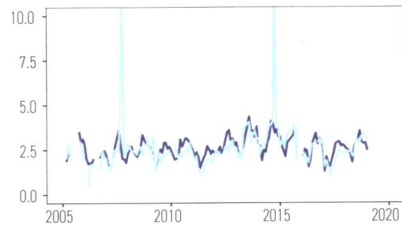	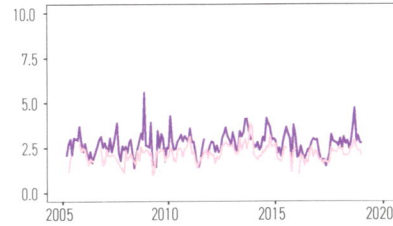	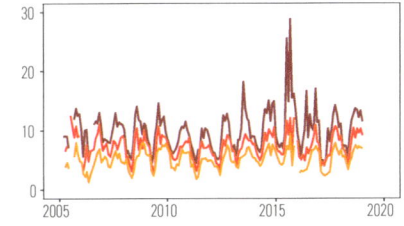
Photic Depth (m)	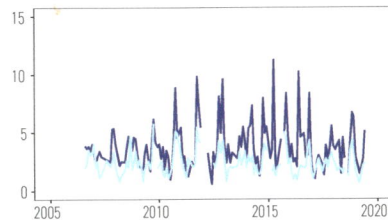	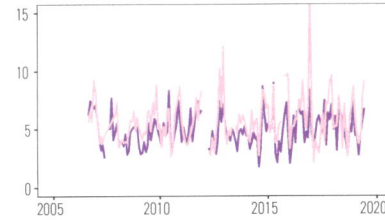	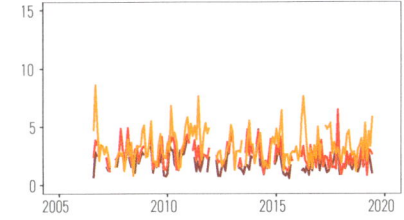
Chlorophyll-a (µg/L)	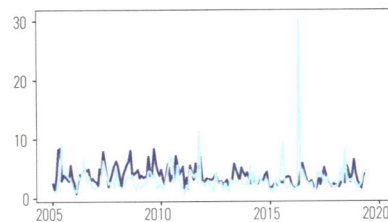	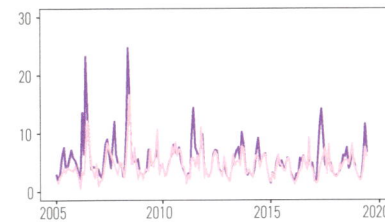	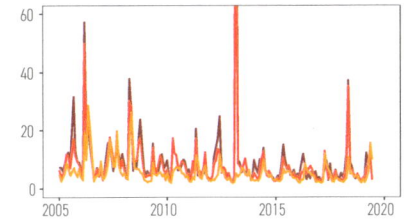

WATER QUALITY UPDATES

75

CEC Tiers

A tiered, risk-based framework guides monitoring and management actions for CECs detected in the Bay. Extensive CEC monitoring by the Regional Monitoring Program has not identified any "high concern" CECs (having a high probability of causing moderate to high impacts on aquatic life). However, there are several CECs that have been placed in the "moderate concern" category (having a high probability of at least a low level impact on Bay aquatic life). Placement in this category flags these contaminants for more intensive monitoring and more aggressive management actions.

Based on findings from recent RMP studies, five CECs have been added to the moderate concern category in the past two years (PFOA and long-chain carboxylate PFAS, imidacloprid, bisphenols, organophosphate esters, and microplastics) and one has been reclassified as a low concern (PBDEs). Key findings for bisphenols and organophosphate esters are presented on the following two pages (**77 and 78**). **PFOA and long-chain carboxylate PFAS** are fluorinated chemicals that are in the range of concentrations that disrupt gene functions in seals and that are not declining in Bay seals or cormorant eggs. **Imidacloprid** is a pesticide that has been detected in Bay water at a level exceeding a protective threshold, and is in widespread and increasing use. **Microplastics** are a highly diverse class of pollutants where protective thresholds have not been established, but technical experts and RMP stakeholders are in consensus that their persistence, increasing abundance, and potential risks to aquatic life merit the moderate concern classification. **PBDE** concentration declines have been documented in many Bay matrices (e.g., **page 81**), to a degree that they present a low risk to humans and aquatic life.

HIGH CONCERN	Moderate or High Impact	None
MODERATE CONCERN	Low Impact	PFOS, PFOA, Long-Chain Carboxylates Fipronil, Imidacloprid Alkylphenols, Alkylphenol Ethoxylates Bisphenols, Organophosphate Esters Microplastics
LOW CONCERN	Limited Impact	PBDEs and HBCD Pyrethroids* Pharmaceuticals Personal Care & Cleaning Products PBDDs / PBDFs
POSSIBLE CONCERN	Uncertainty as to Impact	Alternative Flame Retardants Other PFAS (Fluorinated Chemicals) Pesticides, Plastic Additives Siloxanes, SDPAs, UV-BZTs, others

* Pyrethroids are of low concern in the Bay, but high concern in Bay Area urban creeks

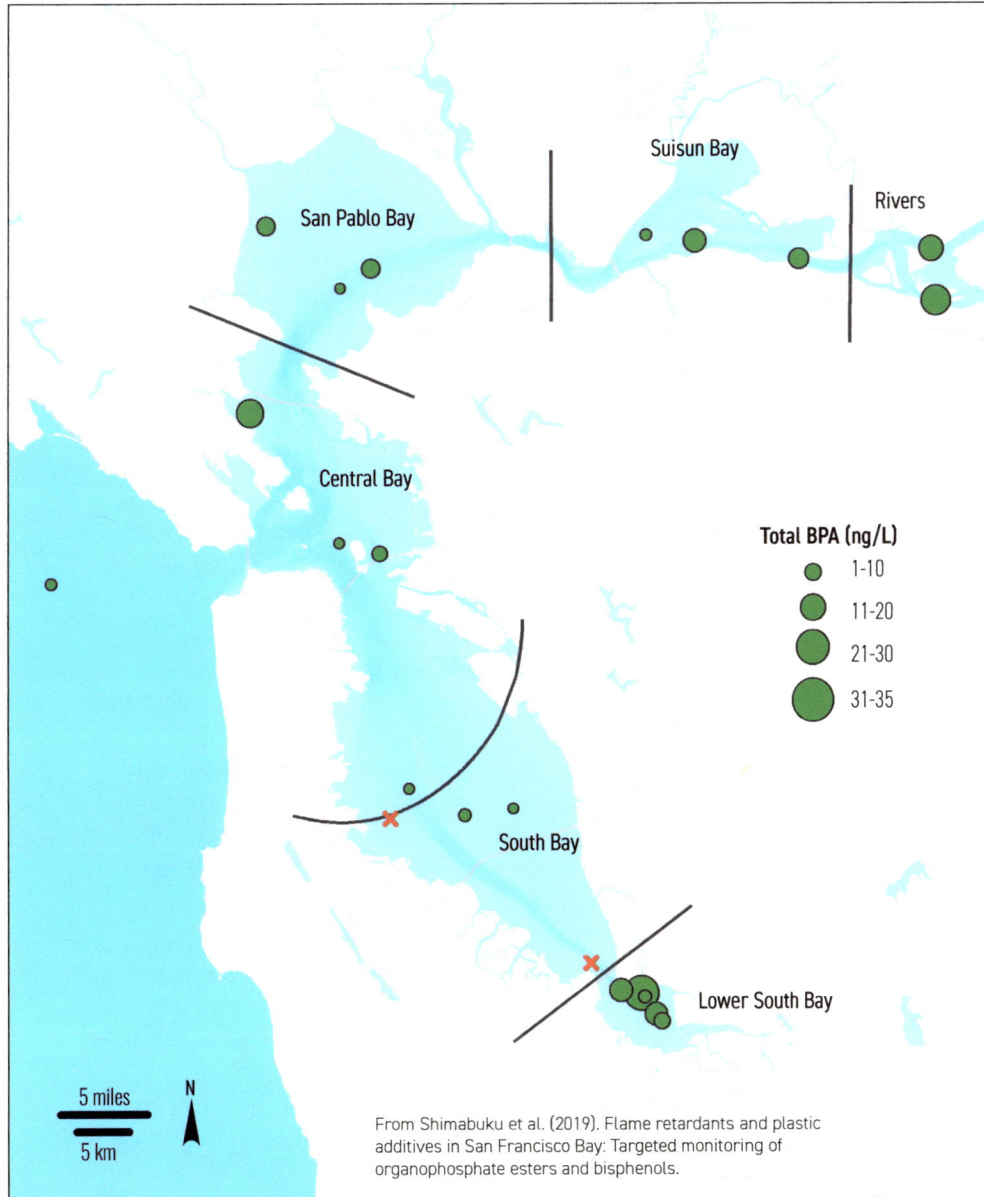

From Shimabuku et al. (2019). Flame retardants and plastic additives in San Francisco Bay: Targeted monitoring of organophosphate esters and bisphenols.

Bisphenols

Bisphenols are an endocrine-disrupting class of synthetic compounds that are manufactured at high volumes, water soluble, and not effectively removed via traditional wastewater treatment processes. Bisphenol A (BPA) and bisphenol S (BPS) are the two most prominent bisphenols, with national production and import volumes in the billions and millions of pounds, respectively. Bisphenols have varying chemical structures and properties, which allow for an array of desirable characteristics (e.g., durability, non-corrosivity, stability) and a plethora of applications in industrial and consumer products. Bisphenols are best known as stabilizing agents and plastic additives in polycarbonate plastic in diverse products such as medical devices, water pipes, baby products, and vehicles. When the use of BPA was banned in baby bottles in 2012, industry began substituting other bisphenols or "alternatives" for BPA. Popular "BPA free" disclaimers on products imply that other BPA alternatives are, presumably, safer alternatives. Though little is known about the toxicity of BPA alternatives, they are structurally similar to BPA and have demonstrated links to the same list of toxic effects through varying metabolic pathways at similar, and sometimes greater, potencies. Production of all bisphenols has increased substantially and is forecast to continue growing.

Monitoring of 16 bisphenols in Bay water was conducted in 2017 at 22 sites during the dry season. Only two bisphenols, BPA and BPS, were detected. Levels of BPA (with a maximum of 35 ng/L in a Lower South Bay sample) were in the range of a 60 ng/L threshold for protection of aquatic life. A threshold for BPS has not yet been established. Bisphenols were determined to merit classification as emerging contaminants of moderate concern for the Bay (**page 76**) due to the presence of individual contaminants in the Bay at levels comparable to or exceeding protective thresholds; the potential for cumulative impacts on endocrine disruption and other toxic effects; the poorly understood spectrum of environmental fates; and the expected increase in production and use.

WATER QUALITY UPDATES

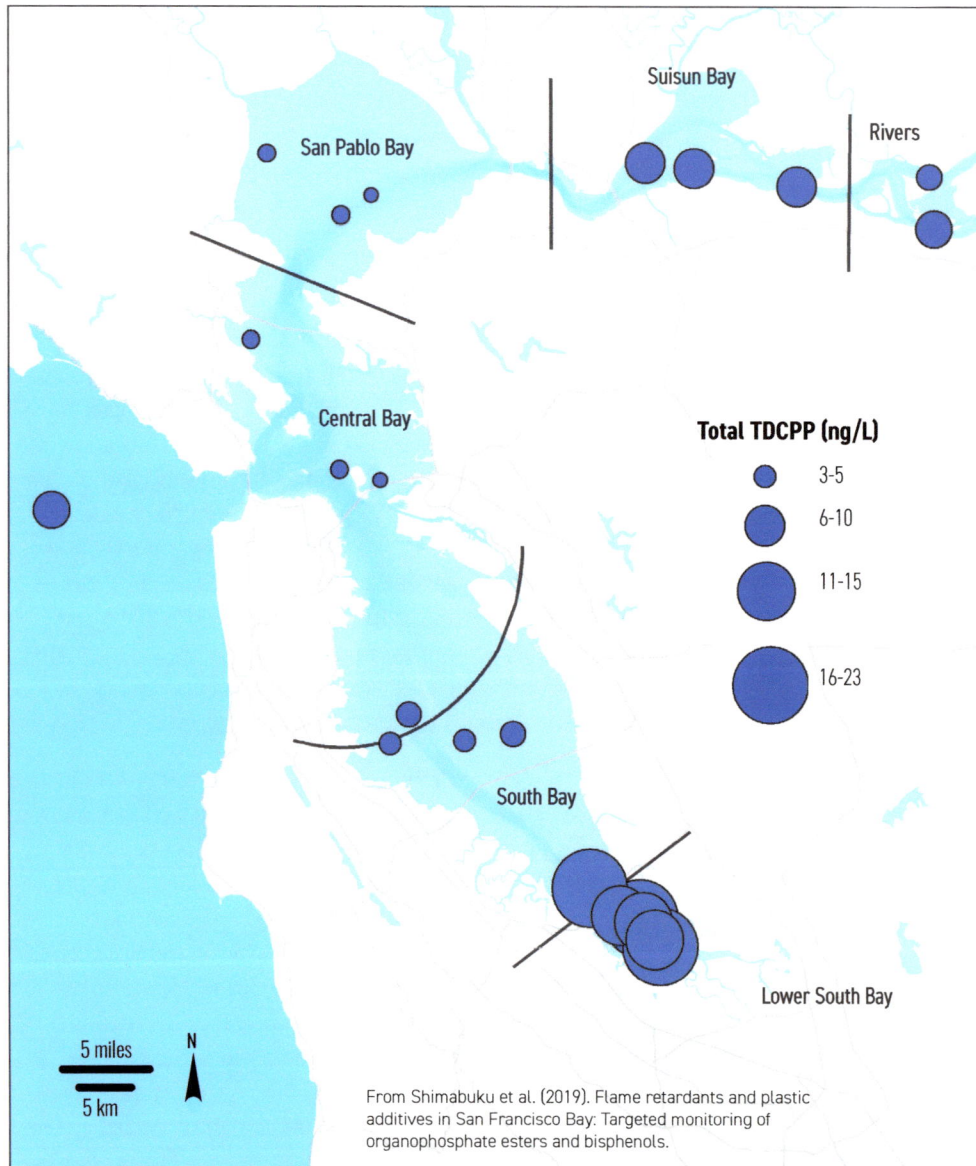

Total TDCPP (ng/L)

○ 3-5

● 6-10

● 11-15

● 16-23

From Shimabuku et al. (2019). Flame retardants and plastic additives in San Francisco Bay: Targeted monitoring of organophosphate esters and bisphenols.

Organophosphate Esters

Organophosphate esters (OPEs), used both as flame retardants and plastic additives, like bisphenols, are an endocrine-disrupting class of synthetic compounds that are manufactured at high volumes, water soluble, and not effectively removed via traditional wastewater treatment processes. When PBDE flame retardants were phased out and banned in the 2000s, OPEs were a popular substitute. In addition to use as flame retardants, OPEs are used as plastic and hydraulic-fluid additives, antifoaming agents, and ingredients in lacquers and floor polishes. Use of OPEs has drastically increased in recent decades and is projected to continue expanding. Though OPE toxicity is not well understood, endocrine-disrupting effects have been demonstrated at environmentally relevant levels. OPEs have also been linked to cancer, neurotoxicity, and adverse effects on fertility. Their industrial popularity, global environmental ubiquity, mobility, toxicity, and, in the case of some OPEs, persistence, give OPEs the potential to cause widespread adverse ecological effects.

Monitoring of 22 OPEs in Bay water was conducted in 2017 at 22 sites in the dry season. Fifteen OPEs were detected in samples from at least one site. Concentrations of one OPE (TDCPP) exceeded a threshold of 20 ng/L for protection of marine life at a few sites (maximum concentration of 23 ng/L; median 6.2 ng/L). OPEs merit classification as emerging contaminants of moderate concern for the Bay, due to the presence of TDCPP in the Bay at levels comparable to or exceeding protective thresholds, the potential for cumulative impacts on endocrine disruption and other toxic effects, the poorly understood spectrum of environmental fates, and the expected increase in production and use.

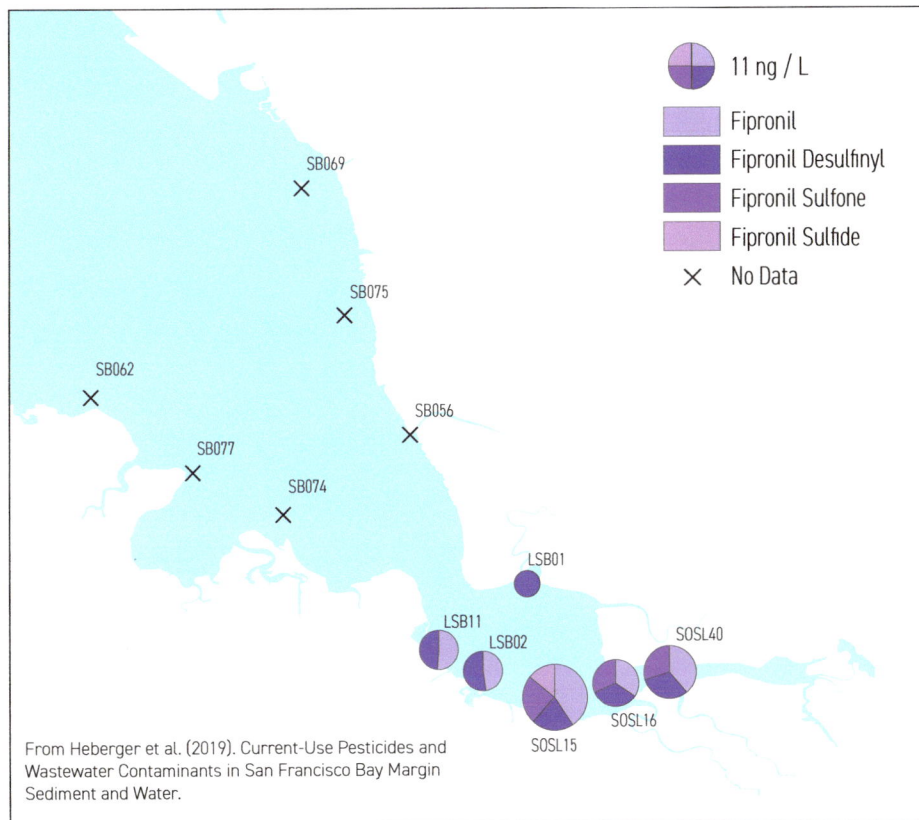

Legend:
- 11 ng / L
- Fipronil
- Fipronil Desulfinyl
- Fipronil Sulfone
- Fipronil Sulfide
- × No Data

From Heberger et al. (2019). Current-Use Pesticides and Wastewater Contaminants in San Francisco Bay Margin Sediment and Water.

Fipronil

Fipronil is an insecticide that has been classified by the RMP as an emerging contaminant of moderate concern for the Bay for the past several years because levels observed in sediment are in the range of toxicity thresholds for freshwater aquatic life (limited toxicity data are available for estuarine or marine species). Fipronil is used for flea, ant, and termite control in California. Outdoor pesticide use can contaminate local creeks and urban runoff that enters the Bay. A recent RMP study on fipronil identified spot-on flea control products as a likely important source of this contaminant to the Bay, as even advanced municipal wastewater treatment facilities are not able to achieve significant removal.

In the summer of 2017, the RMP measured pesticides and other emerging contaminants in water samples from 12 stations in the margins of South Bay. Eighteen pesticides were detected in filtered water samples, and none were detected in suspended sediment. Three of the pesticides – carbendazim, fipronil, and imidacloprid – were detected in some samples at concentrations greater than thresholds for protection of aquatic life in freshwater settings. The highest concentration of fipronil was 12 ng/L, exceeding the USEPA aquatic life benchmark of 11 ng/L for chronic toxicity to freshwater invertebrates. The median concentration of fipronil was 5.1 ng/L. Findings from this study provided additional support for the RMP's designation of fipronil as a moderate concern contaminant. Fipronil degradates are also persistent and toxic, and were detected as well, though at concentrations below available toxicity thresholds.

PFOS in Eggs

Cormorant eggs are a valuable indicator of regional patterns in contamination of the open Bay food web, both for legacy contaminants like mercury and PCBs and emerging contaminants like PFOS. PFOS concentrations in cormorant eggs have been higher in South Bay than in Central Bay (Richmond Bridge) or Suisun Bay (Wheeler Island). South Bay concentrations have varied considerably, falling from over 1200 ppb in 2006 and 2009 to approximately 400 ppb in 2012, then rising to around 600 ppb in 2016, then falling again to 250 ppb in 2018. The concentration in Central Bay in 2018 (27 ppb) was substantially lower than in previous years. The Suisun Bay colony could not be sampled in 2018. PFOS concentrations in cormorant eggs in South Bay may be of concern. Field studies have indicated a 50% reduction in hatching success of tree swallows at a PFOS concentration of 500 ppb wet weight in eggs (Custer et al. 2013), a level similar to that observed in South Bay cormorant eggs.

FOOTNOTE: Average PFOS concentrations (ppb wet weight) in cormorant egg composites. Each point represents three composites, with 7 eggs in each composite. The Suisun Bay colony could not be sampled in 2018.

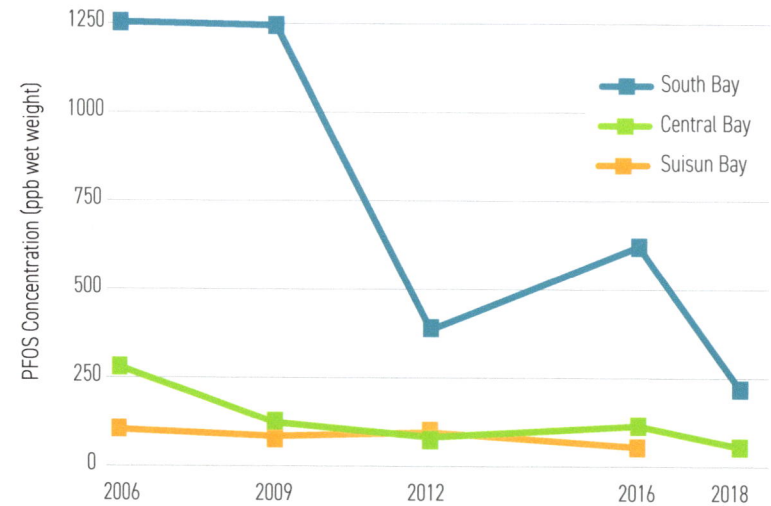

◄ **Double-crested Cormorants in flight**
(Shira Bezalel, SFEI)

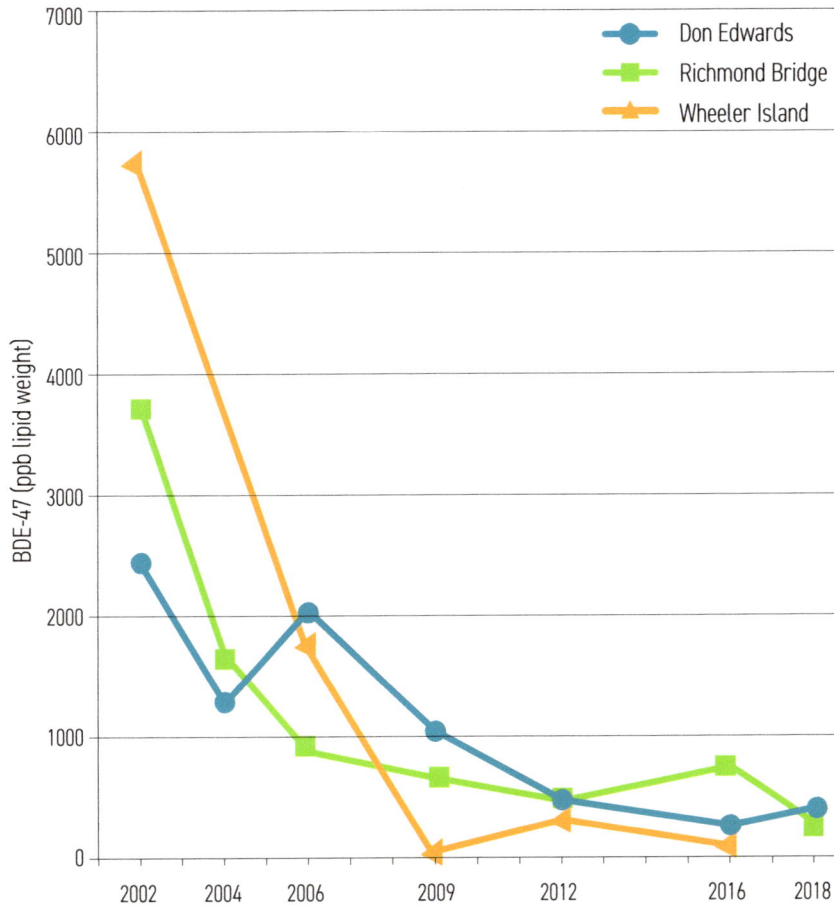

BDE-47 (ppb lipid weight) — vertical axis

Legend:
- Don Edwards
- Richmond Bridge
- Wheeler Island

FOOTNOTE: Average BDE-47 concentrations (ppb lipid weight) in cormorant egg composites. BDE-47 is the most abundant BDE congener in the eggs and is presented as an index of PBDEs as a whole. Each point represents three composites, with 7 eggs in each composite. The Suisun Bay colony could not be sampled in 2018.

PBDEs in Eggs

Cormorant eggs are one of the indicators of a marked decline in PBDE concentrations in the Bay since phase-outs and bans occurred in the mid-2000s. These actions resulted in a rapid response, with concentrations falling to much lower levels by 2009. The decline was fastest in North Bay (Wheeler Island) and slowest in South Bay (Don Edwards). In 2017, due to the declines in multiple indicators (bird eggs, bivalves, sport fish, and sediment) and resolved uncertainties about risks to humans and wildlife, PBDEs were reclassified by the RMP from a moderate concern to a low concern for the Bay.

Mercury in Sediment

Mercury binds to sediment particles, so mercury concentrations in the sediment deposits on the bottom of the Bay are an important index of contamination of the ecosystem. The RMP measures mercury and other pollutants in sediment across the entire Bay once every four years, most recently in 2018.

In 2015, the RMP began an additional set of surveys of sediment in the margins of the Bay (shallow areas that had previously not been monitored), beginning with Central Bay. Mercury concentrations in margin sediment were very similar to concentrations in the deeper waters of Central Bay. However, a handful of sites with relatively high concentrations (above 0.5 ppm) were observed in the margins. In 2017, South Bay margins were sampled. Average mercury concentrations in South Bay margin sediment were actually significantly lower than in the open waters of South Bay (**facing page, lower graph**) which is counterintuitive because the margins are generally closer to pollution sources. When the concentrations were adjusted for the amount of fine-grained sediment in the samples, however, there was no difference between margins and open Bay. Mercury concentrations are higher on fine-grained sediment particles due to their higher ratio of surface area to volume.

No trends in mercury concentrations are evident in long-term data for the open Bay across the segments since the current sampling design was established in 2001 (facing page, upper graphs). Average concentrations in 2018 in San Pablo Bay (0.35 ppm) and South Bay (0.30 ppm) were the highest yet observed for these segments.

FOOTNOTE: Points on the map show all available dry season RMP data from 2002-2018 (circles) along with Central Bay and South Bay margin data (triangles) from 2015 and 2017, respectively.

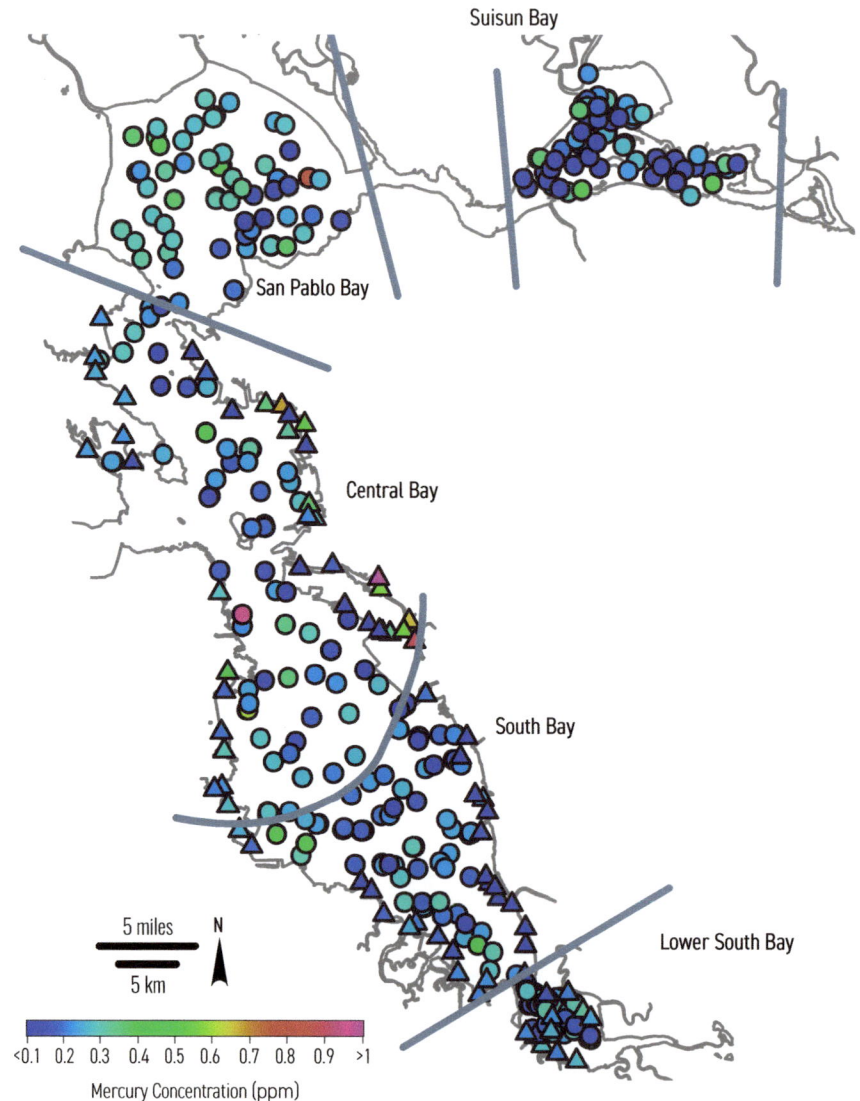

Mercury Sediment Trends

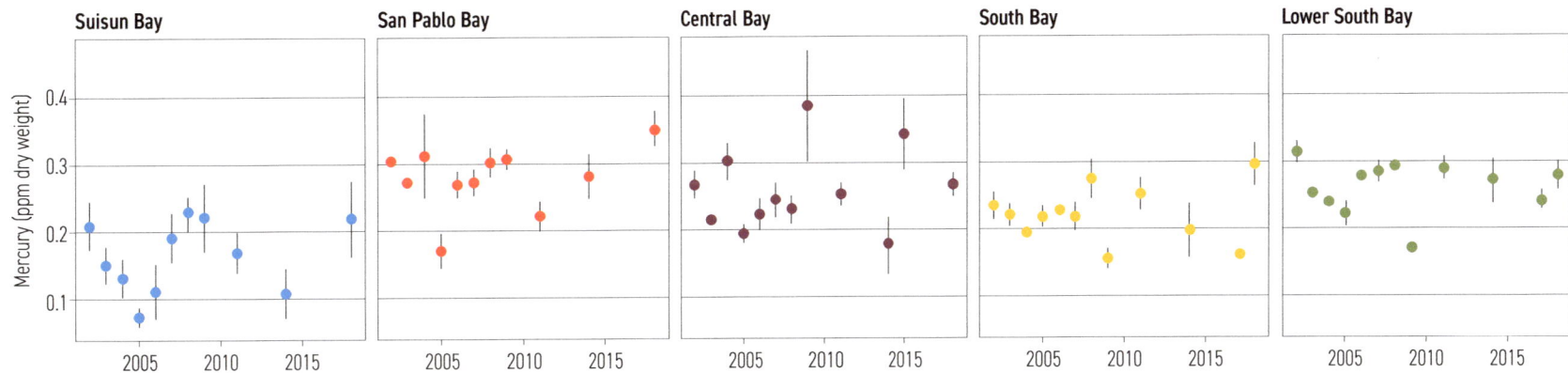

Suisun Bay — **San Pablo Bay** — **Central Bay** — **South Bay** — **Lower South Bay**

Mercury (ppm dry weight)

FOOTNOTE: Averages plus or minus one standard error for all available dry season RMP data from 2002-2018 along with Central Bay and South Bay margin data from 2015 and 2017, respectively.

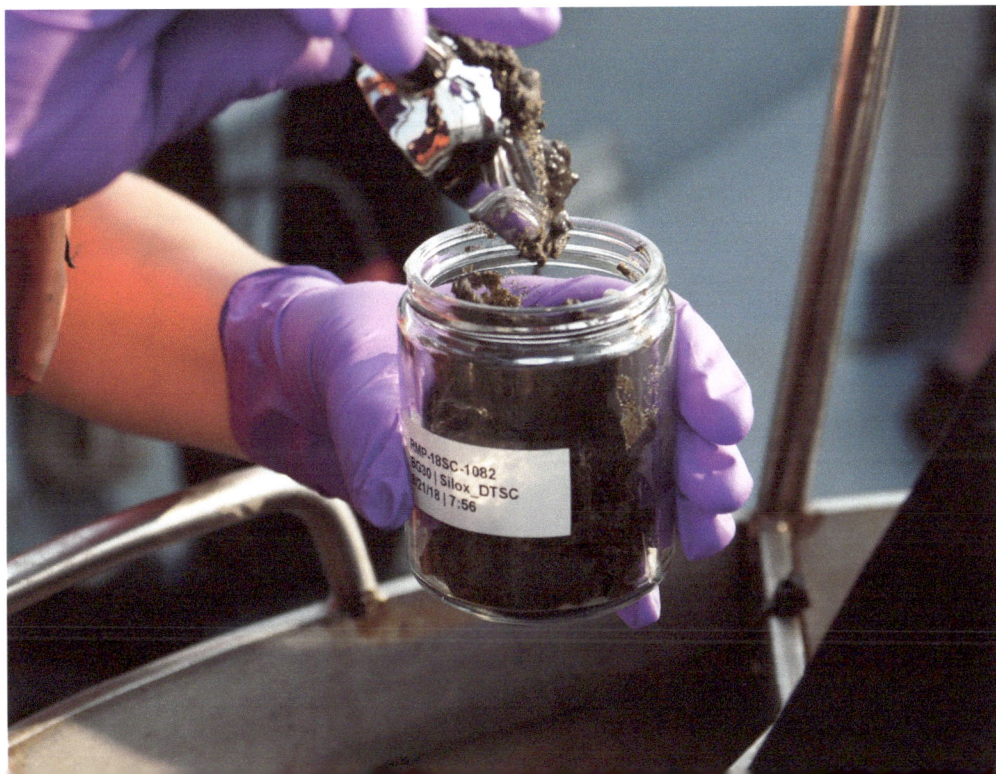

South Bay and Lower South Bay

Mercury (ppm dry weight)

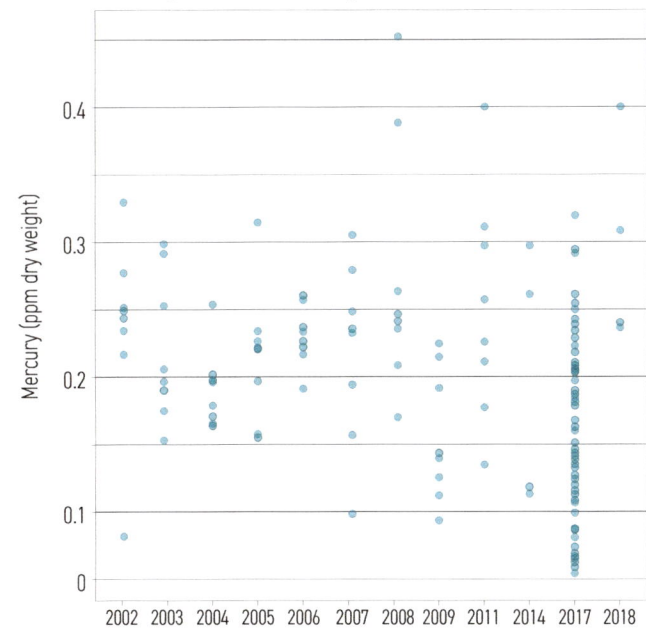

◀ **Placing Bay sediment in a sample jar** (Shira Bezalel, SFEI)

GRAPH DETAILS ON PAGE 96

Water Flow and Mercury Load from the Guadalupe River

Efforts to reduce mercury loads to the Bay are primarily focusing on the Guadalupe River and urban stormwater. The Guadalupe River carries runoff from the New Almaden Mercury Mining District, historically the nation's largest mercury mining region and a continuing source of legacy contamination to Lower South Bay. Load reduction activities in the Guadalupe watershed have been underway for over a decade and are planned to continue for at least another two decades.

Guadalupe River flow has a major influence on mercury loading to the Bay, and the flow in the wet season of 2016/2017 was extremely high. A series of large storms yielded an estimated total flow for the water year of 249 million cubic meters, the highest annual flow observed since records began in 1932. The flow in 2017/2018 (46 million cubic meters) was relatively low, below the average for the most recent 30-year period (66 million cubic meters). Flow was above average again in 2018/2019, with a preliminary estimate of 129 million cubic meters.

The RMP was able to sample mercury in the Guadalupe during the record high flows of the 2016/2017 wet season, adding to a relatively extensive long-term dataset for loading from this watershed. The estimated mercury load for that wet season (1072 kg) was far greater than the sum of the loads from 2003-2016 (231 kg). An estimated load of 3 kg in 2018 was then followed by another large estimated load (301 kg) in the above-average rainfall year of 2019. These estimates highlight the highly episodic nature of mercury transport from the watershed, which poses challenges for both monitoring and management.

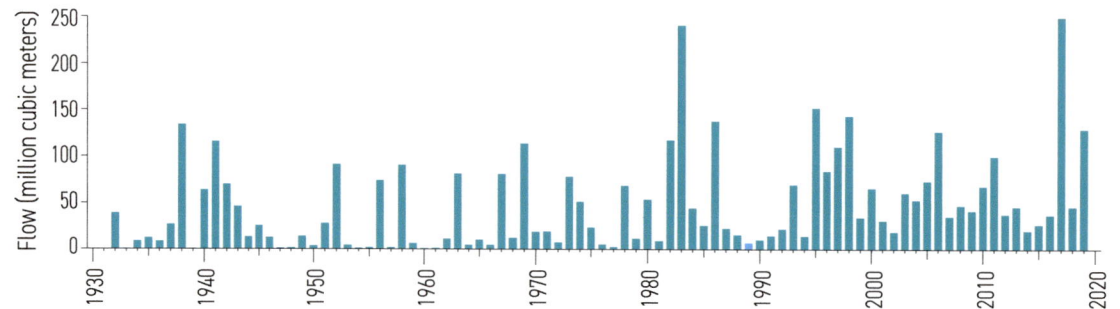

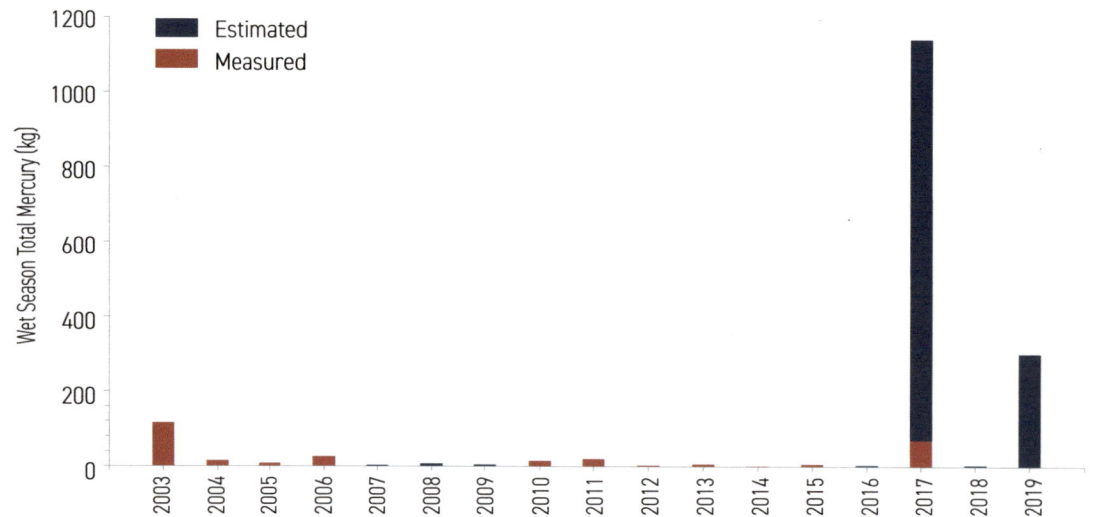

MERCURY

Mercury in Eggs

The RMP tracks concentrations of mercury and other pollutants in cormorant eggs as another means of assessing trends in contamination of the food web over time. The period of record now spans 15 years or more at three locations in Suisun Bay (Wheeler Island), Central Bay (Richmond Bridge), and South Bay (Don Edwards National Wildlife Refuge). Mercury concentrations have been highest, and most variable, in the South Bay. No long-term trend is apparent in these data.

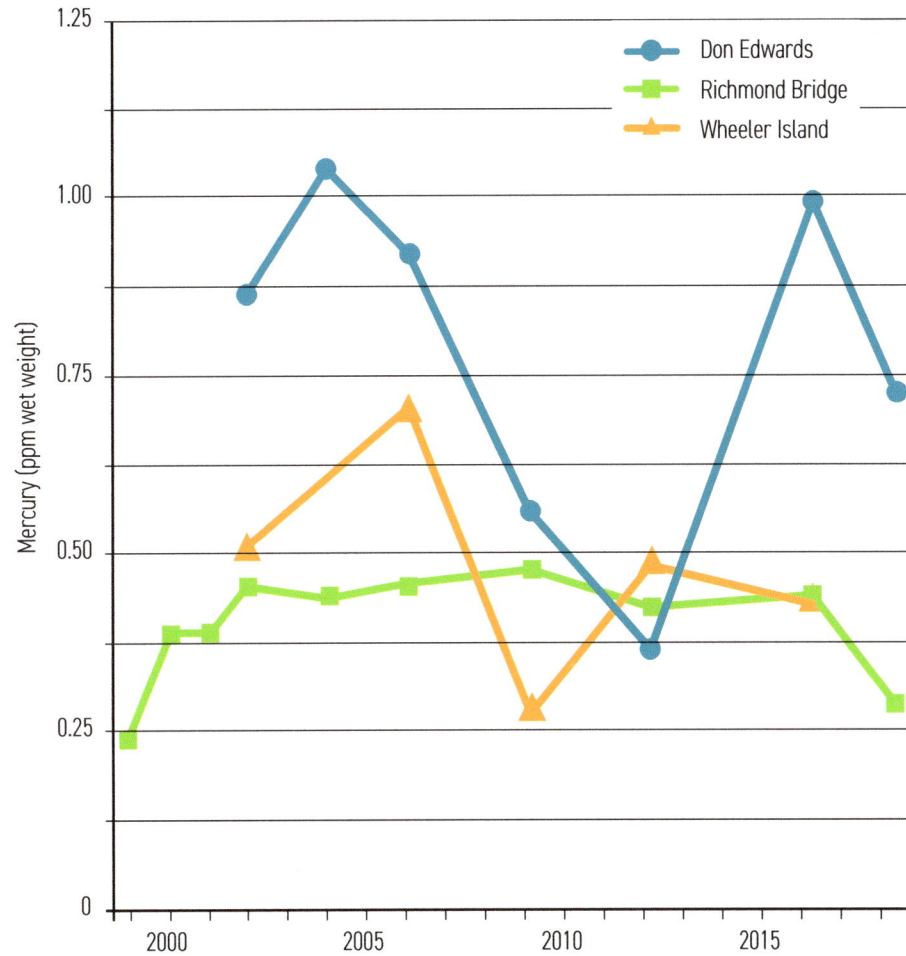

FOOTNOTE: Average mercury concentrations (ppm wet weight) in cormorant egg composites. Each point represents three composites, with 7 eggs in each composite. The Suisun Bay colony could not be sampled in 2018.

PCBs in Sediment

Like mercury, PCBs bind to sediment particles, so PCB concentrations in the sediment deposits on the bottom of the Bay are an important index of contamination of the ecosystem. The RMP measures PCBs and other pollutants in sediment across the entire Bay once every four years, most recently in 2018.

In 2015, the RMP began an additional set of sediment surveys in the margins of the Bay (shallow areas that had previously not been monitored), beginning with Central Bay. PCB concentrations at many sites in the Central Bay margins were higher than the maximum concentration (40 ppb) observed at deeper water sites in Central Bay. Although the median concentration of PCBs in the margins (13 ppb) was similar to the median for the open Bay (11 ppb), the 75th percentile for the margins (32 ppb) was twice as high as the 75th percentile for the open Bay (16 ppb).

In 2017, South Bay margins were sampled. Average PCB concentrations in South Bay margin sediment (11.5 ppb) were slightly, but statistically significantly, higher than in the open waters of South Bay (10.3 ppb). The difference was larger when the concentrations were adjusted for the amount of fine-grained sediment in the samples (17.6 ppb versus 14.3 ppb). PCB concentrations are higher on fine-grained sediment particles due to their higher ratio of surface area to volume.

No trends in sediment PCB concentrations are evident in the long-term data for the open Bay across the segments since the current sampling design was established in 2001. Concentrations within each segment have been quite consistent over the years, and the 2018 averages were within the range of previous observations.

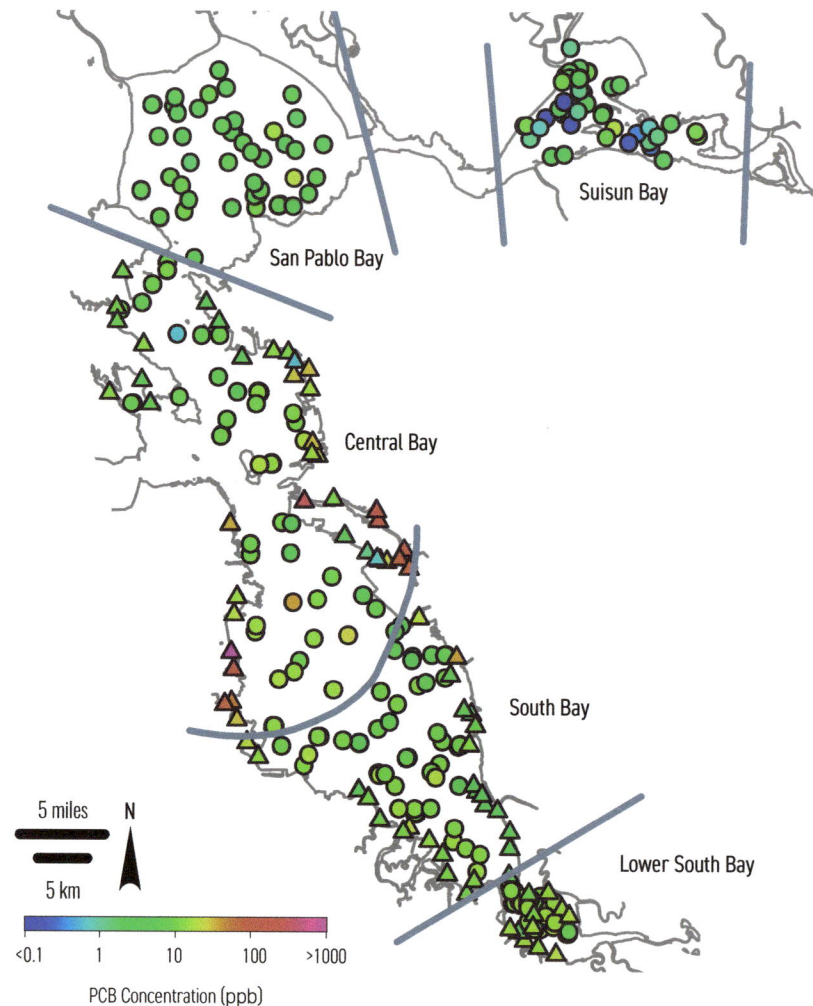

FOOTNOTE: Points on the map show all available dry season RMP data from 2002-2018 (circles) along with Central Bay and South Bay margin data (triangles) from 2015 and 2017, respectively.

PCBs

PCBs in Eggs

The RMP tracks concentrations of PCBs and other pollutants in cormorant eggs as another means of assessing trends in contamination of the food web over time. The period of record now spans 15 years or more at three locations in Suisun Bay (Wheeler Island), Central Bay (Richmond Bridge), and South Bay (Don Edwards National Wildlife Refuge). Average PCB concentrations have been higher in South Bay and Central Bay than in Suisun Bay. The average concentration in South Bay in 2015 was the lowest yet measured for that region. No distinct long-term trend is apparent in these data.

FOOTNOTE: Average PCB concentrations (ppm lipid weight) in cormorant egg composites. Each point represents three composites, with 7 eggs in each composite. The Suisun Bay colony could not be sampled in 2018.

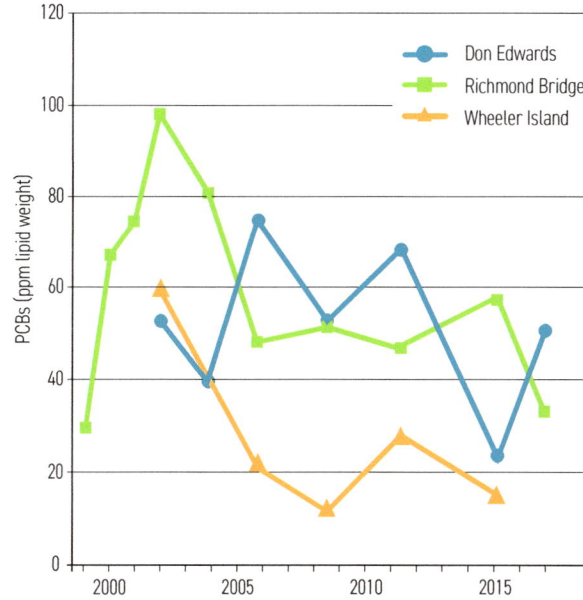

PCBs in Sediment (continued)

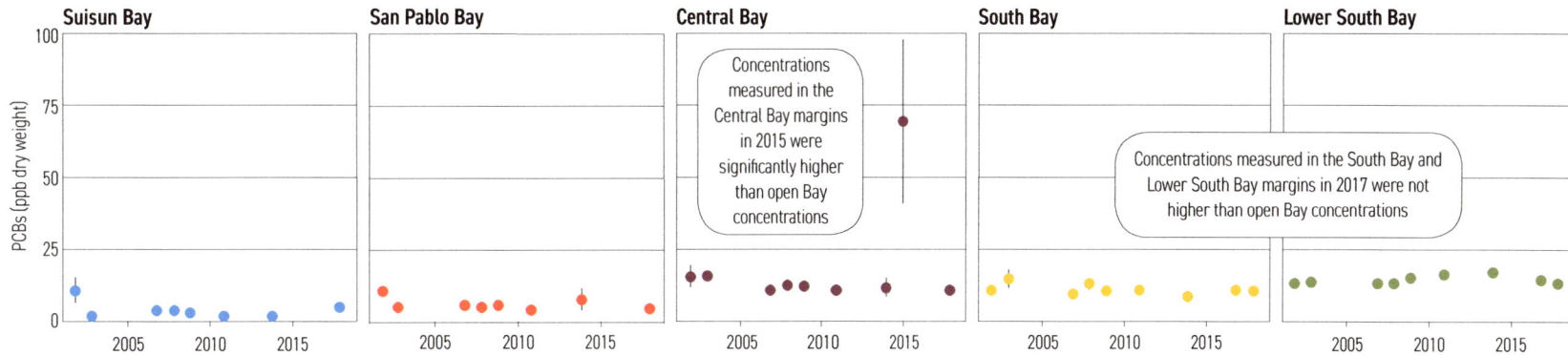

Suisun Bay   **San Pablo Bay**   **Central Bay**   **South Bay**   **Lower South Bay**

Concentrations measured in the Central Bay margins in 2015 were significantly higher than open Bay concentrations

Concentrations measured in the South Bay and Lower South Bay margins in 2017 were not higher than open Bay concentrations

FOOTNOTE: Averages plus or minus one standard error for all available dry season RMP data from 2002-2018 along with Central Bay and South Bay margin data from 2015 and 2017, respectively.

Selenium in Sturgeon

White sturgeon, a species that preys on clams and other bottom-dwelling invertebrates, is recognized as a key indicator of selenium impairment in the North Bay due to its susceptibility and sensitivity to selenium bioaccumulation.

In recent years, the RMP has focused on improving information on impairment through more extensive monitoring of white sturgeon. Non-lethal sampling of muscle plugs from sturgeon, in collaboration with the California Department of Fish and Wildlife, began in 2014 and is greatly expanding this critical dataset.

The long-term dataset for selenium in sturgeon muscle generated by the RMP and other programs suggests that concentrations were relatively high in 1989 and 1990, and fairly constant in subsequent years through 2014. A target of 11.3 ppm in white sturgeon muscle was established in the TMDL for selenium in the North Bay that was approved in 2016. Recent results through 2014 indicate that average concentrations were below the target, but a few samples exceeded it.

More intensive monitoring of selenium in sturgeon muscle plugs was performed in 2015, 2016, and 2017. Sampling during 2015 and 2016 occurred during the last two years of a five-year drought. Long-term monitoring of North Bay clams has shown that dry years are associated with high selenium concentrations, and wet years with low concentrations. Selenium concentrations in sturgeon muscle in 2015 and 2016 were high relative to prior data, with medians and averages near the TMDL target of 11.3 ppm. Sampling in 2017, however, followed a very wet winter, and concentrations in sturgeon were much lower (average of 7.3 ppm). This three-year dataset with relatively large numbers of samples illustrated the influence of water year type on selenium in the North Bay food web. The RMP is continuing to monitor selenium in sturgeon muscle plugs on a biennial basis, with the next round of sampling in fall of 2019.

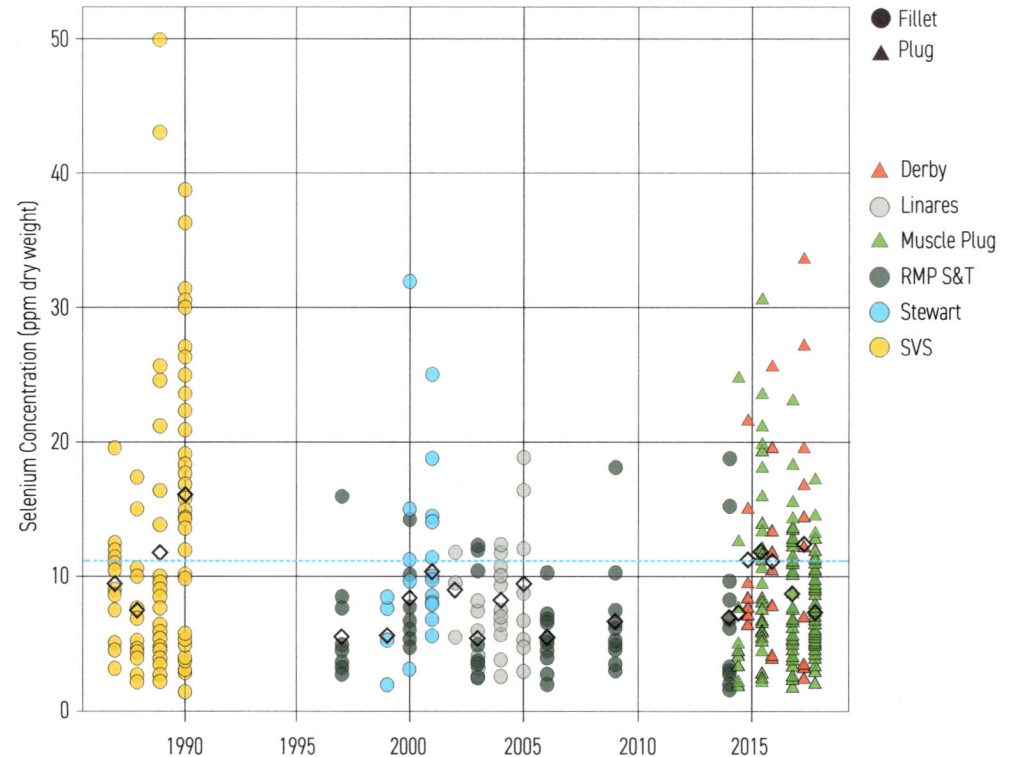

FOOTNOTE: Points represent samples of individual white sturgeon. Mean concentrations for each study, or each year of multi-year studies, are shown in black diamonds. Horizontal blue line indicates the North Bay TMDL target for selenium in sturgeon muscle tissue (11.3 µg/g dw). Data from the RMP and other sources as follows: Derby – Sun et al. (2019); Linares – Casenave et al. (2015); Muscle Plug – Sun et al. (2019b); RMP S&T (1997- 2014); Stewart – Stewart et al. 2004; SVS (Selenium Verification Study) – Urquhart et al. 1991.

SELENIUM

Selenium in Sturgeon (continued)

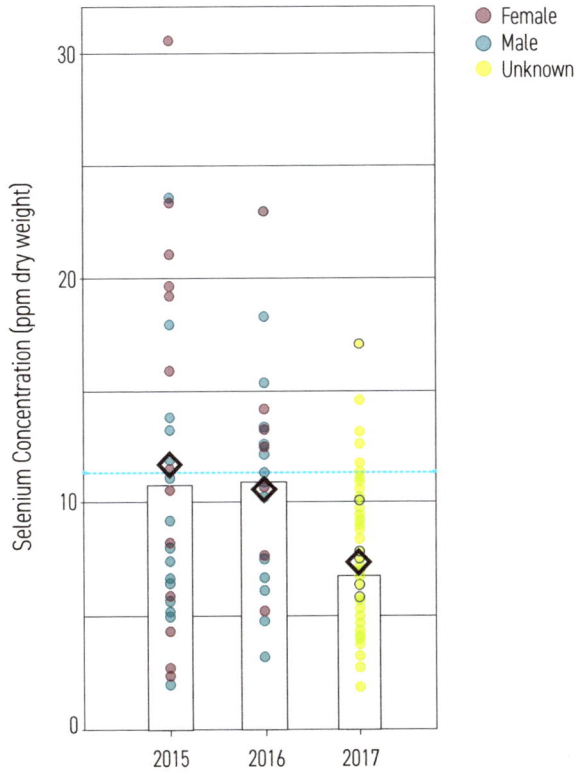

FOOTNOTE: Each point represents an individual sturgeon. Median concentrations are shown as white bars and average concentrations as black diamonds. The horizontal blue line indicates the North Bay TMDL target for selenium in sturgeon muscle tissue (11.3 µg/g dw).

Selenium in Eggs

Avian predators of fish and aquatic invertebrates can also be at risk from selenium accumulation, and avian eggs are therefore another valuable indicator of potential impairment and trends. A selenium standard of 12.5 ppm in bird eggs was approved for Great Salt Lake in 2011. The RMP has tracked selenium concentrations in double-crested cormorant eggs at three locations for a span of up to 19 years. The highest concentration measured in a single composite sample was 8.7 ppm in 2009. Concentrations were unusually high in 2009, and relatively constant in the other years sampled.

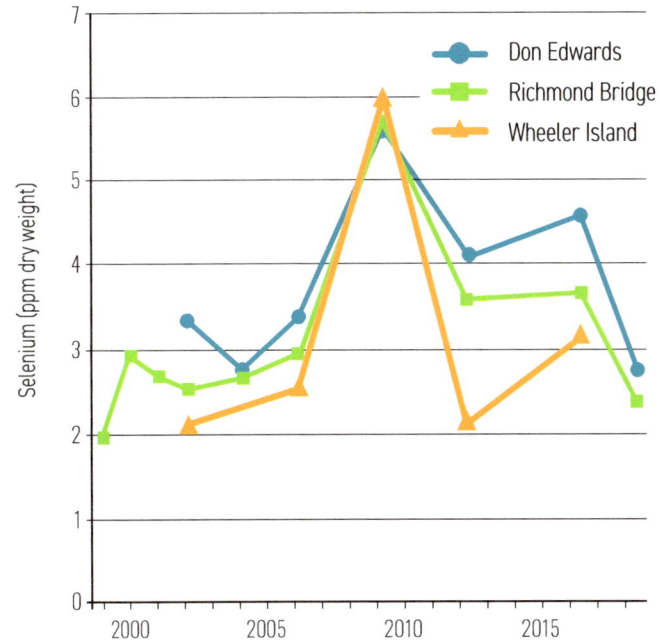

FOOTNOTE: Average selenium concentrations (ppm dry weight) in cormorant egg composites. Each point represents the average of three composites, with 7 eggs in each composite. The Suisun Bay colony could not be sampled in 2018.

WATER QUALITY UPDATES

89

Beach Report Card Summary

Pathogenic organisms found in waste from humans and other warm-blooded animals can pose health risks to people who recreate in contaminated waters. Six Bay beaches are on the 303(d) List of impaired water bodies because fecal indicator bacteria exceed water quality standards, and a TMDL was approved in February 2017 to address this impairment. Another TMDL for two more Bay beaches (Kiteboard Beach and Oyster Point Beach) is in development.

County public health and other agencies routinely monitor fecal indicator bacteria (FIB) concentrations at 26 Bay beaches where water contact recreation is common and provide warnings to the public when concentrations exceed the standards. Using these data, Heal the Bay, a Santa Monica-based non-profit, provides evaluations of over 400 California bathing beaches in Beach Report Cards as a guide to aid beach users' decisions concerning water contact recreation (Heal the Bay 2019). The Report Cards use a familiar A through F grading scale to summarize the results of the county monitoring. Heal the Bay's grading system takes into consideration the magnitude and frequency of exceedances above allowed bacterial levels over the course of the specified time period. The risk of illness from pathogen exposure increases with lower grades.

The Bay-wide average summer grade for 2018 was an A- (GPA of 3.51). The Bay-wide average summer grade has been fairly constant at this level over the past seven years.

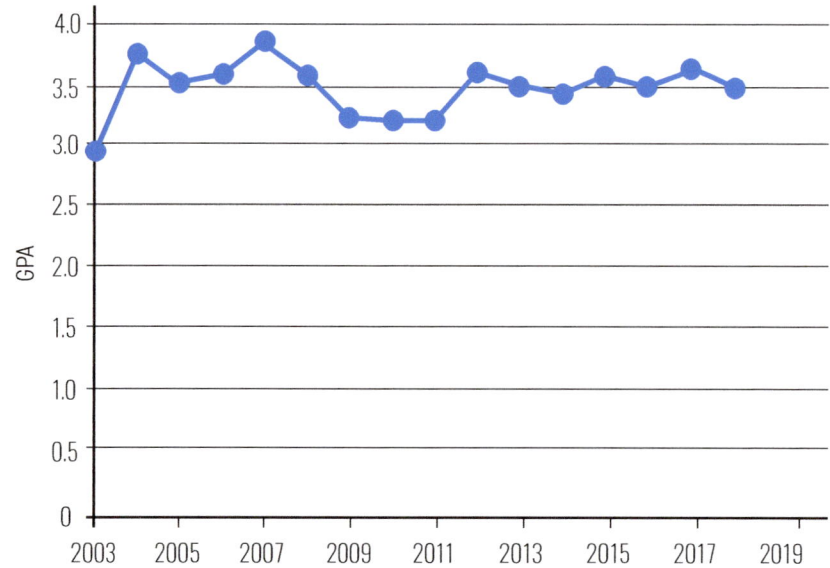

FOOTNOTE: Data from Heal the Bay (2019) and earlier Heal the Bay reports.

Crown Beach, Alameda ▶
(Shira Bezalel, SFEI)

Beach Report Card Details

Overall, the monitoring data and resulting Beach Report Card grades indicate that most Bay beaches are safe for summer swimming, but that bacterial contamination is a concern at a few beaches in the summer, and at a higher number of beaches during wet weather.

Data for the summer beach season in 2018 are available for 26 beaches. In 2018, 17 of the 26 monitored beaches received an A or A+ grade, reflecting minimal exceedance of standards. Seven of these beaches received an A+: Coyote Point, Crissy Field Beach West, Crissy Field Beach East, Aquatic Park Beach 211 Station, Baker Beach Horseshoe Cove NW, Schoonmaker Beach, and Paradise Cove. Nine of the 26 beaches monitored in the summer in 2018 had grades of B or lower, indicating varying degrees of exceedance of bacteria standards. Aquatic Park in San Mateo County, Crown Beach Crab Cove in Alameda County, and Keller Beach South in Contra Costa County received a D. These low grades indicate an increased risk of illness or infection. Overall, the average grade for the 26 beaches monitored from April-October was an A-.

During wet weather, which mostly occurs in the winter, water contact recreation is less popular but is still enjoyed by a significant number of Bay Area residents. Bacteria concentrations are considerably higher during wet weather due to stormwater runoff and sewer overflows, making the Bay less safe for swimming. In wet weather, 10 of 27 beaches with data (37%) had F grades. The following 10 beaches had grades of F: Lakeshore Park in San Mateo County; Crown Beach Crab Cove, Crown Beach Shoreline Drive, and Crown Beach Bird Sanctuary in Alameda County; Islais Landing and Candlestick Point Windsurfer Circle in San Francisco County; and Fort Baker Horseshoe Bay NW, Schoonmaker Beach, China Camp, and McNears Beach in Marin County. Only eight of the beaches (30%) had grades of A or A+ in wet weather. The overall average GPA for these 27 beaches in wet weather was 2.12 (a C).

Legend:
- A (blue)
- B (green)
- C (yellow)
- D (orange)
- F (red)

Beach Report Card Grades (2011–2018)

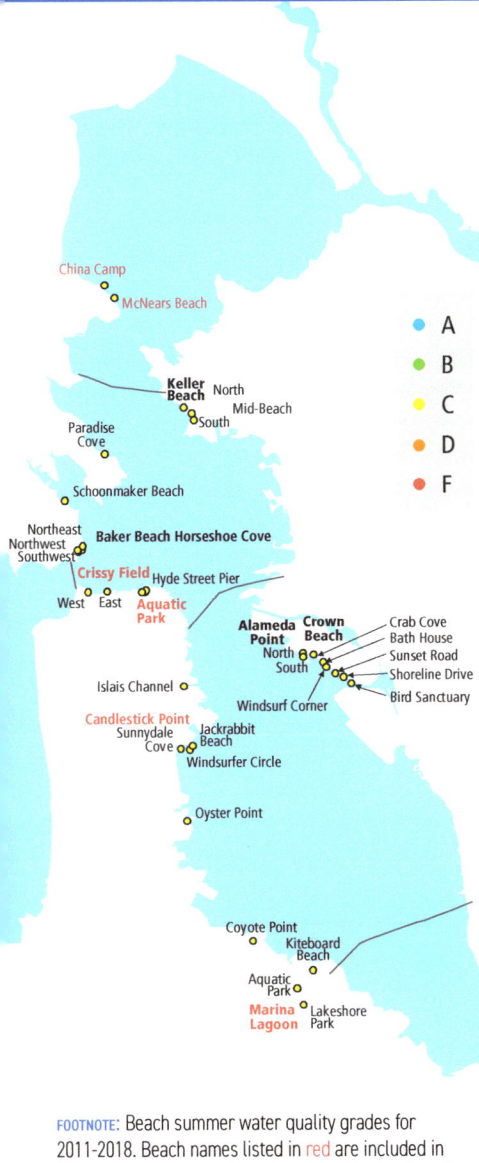

	2011	2012	2013	2014	2015	2016	2017	2018
SAN MATEO COUNTY								
Oyster Point	A		A	A		A		A
Coyote Point	A	A	A	A	A		A	A
Marina Lagoon Aquatic Park	D	D	D	D	C	B	B	D
Lakeshore Park	F	D	D	F		D	F	B
Kiteboard Beach		A	C		A			
ALAMEDA COUNTY								
Alameda Point North			A	A	A	A	A	A
South			A	A	A	A	A	A
Crown Beach Crab Cove			A	A	A	B	B	D
Bath House			A	A	A	A	A	A
Windsurf Corner			A	A	A	A	A	A
Sunset Road			A	A	A	A	A	A
Shoreline Drive			A	A	A	A	A	A
Bird Sanctuary	B	A	A	A	A	A	A	A
CONTRA COSTA COUNTY								
Keller Beach North	B	A	A	A	C	A	B	B
Mid-Beach	B							
South	B	A	A	A	C	A	B	D
SAN FRANCISCO COUNTY								
Crissy Field Beach West	A	A	A	A	A	A	A	A
Mid-Beach		A	A	A	A	A	A	A
East		A	A	A	A	A	A	A
Aquatic Park Beach 211 Station	B	A	A	A	A	A	A	A
Hyde Street Pier	A	A	A	A	A	A	A	A
Islais Channel	A					A	A	B
Candlestick Point Jackrabbit Beach	B	A						
Windsurfer Circle	A	C	A	C	C	A	A	A
Sunnydale Cove	A	A	A	F	D	A	A	A
MARIN COUNTY								
Baker Beach Horseshoe Cove NE	A	A	A	A	A	A	A	A
NW	A	A	A	A	A	A	A	A
SW	A	A	A	A	A	A	A	A
Schoonmaker Beach	B	A	A	A	A	A	A	A
Paradise Cove		A	A	A	A	A	A	A
China Camp		A	A	A	A	A	A	A
McNears Beach			A	A	A	A	A	A

FOOTNOTE: Beach summer water quality grades for 2011-2018. Beach names listed in red are included in the Bay Beaches TMDL. Data from Heal the Bay (2019) and previous Heal the Bay reports.

WATER QUALITY UPDATES

Copper in Water

Copper in the Bay was a major concern in the 1990s. An evaluation of the issue by the Water Board and stakeholders, based on an extensive dataset provided by the RMP and other studies showing that most of the copper in the Bay is bound up in a harmless form, concluded that the existing water quality objectives were inappropriately low. These findings led to new Bay-specific water quality objectives for copper (less stringent but still considered fully protective of aquatic life), pollution prevention and monitoring activities to make sure concentrations remain below the objectives, and the 2002 removal of copper from the 303(d) List of pollutants of concern in the Bay.

In order to ensure that concentrations have not increased, monitoring data collected by the RMP are compared to specific trigger levels. If the trigger concentration is exceeded in any Bay segment, the Water Board will investigate causes of the exceedance and consider potential control options. Concentrations in the most recent assessment period were below the triggers (lower right). The Bay-wide mean in 2017 (1.61 µg/L) was the third lowest recorded over the period of record.

To maintain water quality in the Bay, municipalities are required to implement actions to control discharges to storm drains from architectural (e.g., roofs) and industrial (e.g., metal plating) uses of copper, as well as copper used as an algaecide in pools, spas, and fountains. They are also required to address vehicle brake pads, the largest source of copper to the Bay, which they have done through participation in the Brake Pad Partnership, a public-private collaboration whose work led to the passage of legislation (SB 346) requiring that the amount of copper in brake pads sold in California be reduced to no more than 0.5% by 2025.

FOOTNOTE: Points on the map show results from samples collected in 2017; color contours are based on all available RMP data since 2001. Bay-wide trend plot shows annual random-station means with error bars indicating the 95% confidence intervals of the means.

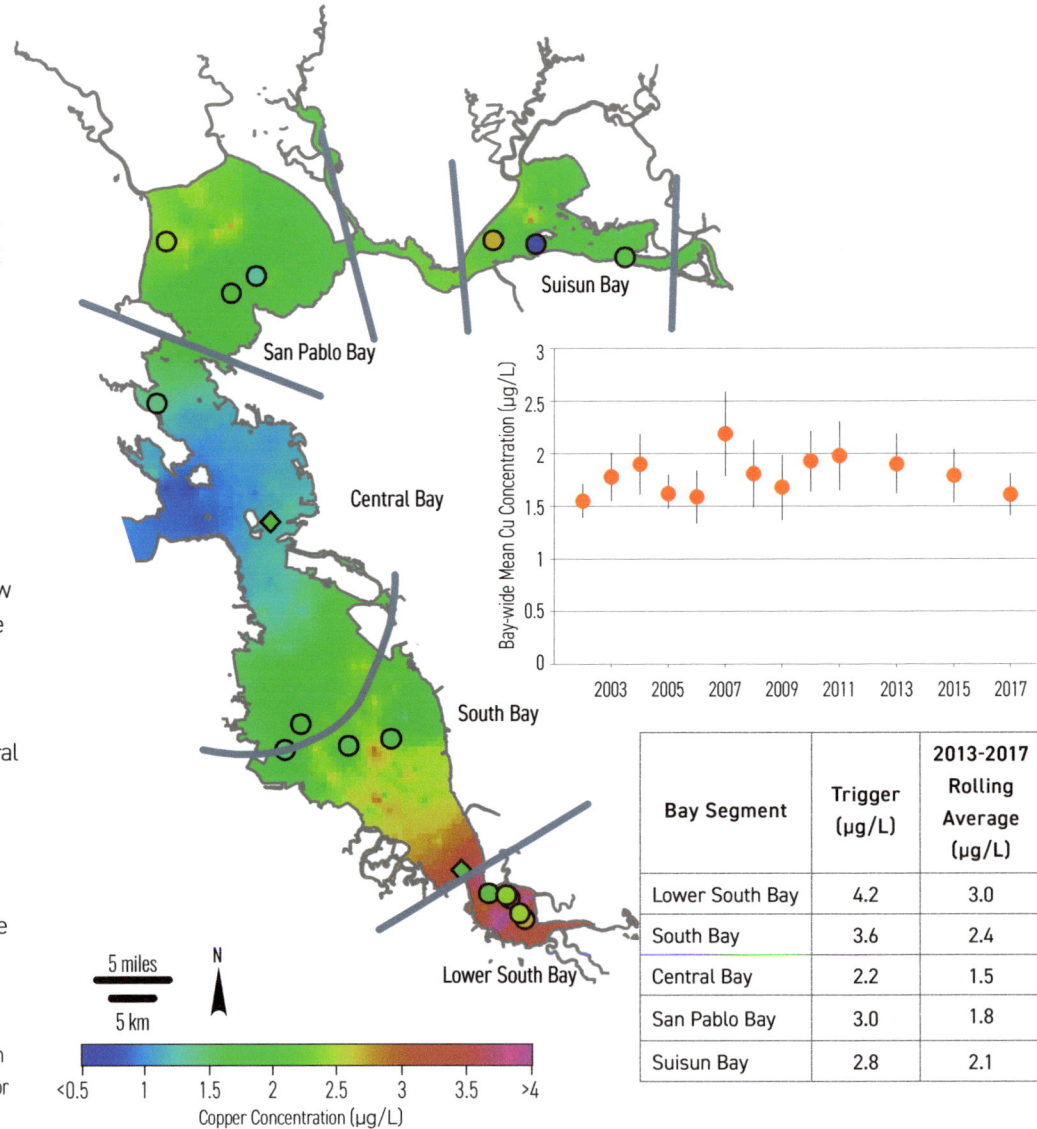

Bay Segment	Trigger (µg/L)	2013-2017 Rolling Average (µg/L)
Lower South Bay	4.2	3.0
South Bay	3.6	2.4
Central Bay	2.2	1.5
San Pablo Bay	3.0	1.8
Suisun Bay	2.8	2.1

Section 303(d) of the 1972 Federal Clean Water Act requires that states develop a list of water bodies that do not meet water quality standards and develop action plans, called Total Maximum Daily Loads (TMDLs), to improve water quality.

The list of impaired water bodies is periodically updated. The RMP is one of many entities that provide data to the State Water Board to assess water quality and inform the 303(d) List. The process for developing the 303(d) List for the Bay includes the following steps:

- development of a draft list of recommendations by the San Francisco Bay Regional Water Board;

- adoption by the State Water Board; and

- approval by USEPA.

The primary pollutants/stressors for the Bay and its major tributaries on the 303(d) List include:

Trace elements: Mercury and Selenium

Pesticides: Dieldrin, Chlordane, and DDT

Other chlorinated compounds:
PCBs
Dioxin and Furan Compounds

Others: Exotic Species, Trash, Polycyclic Aromatic Hydrocarbons (PAHs), and Indicator Bacteria

PARAMETER	STATUS
Copper	Site-specific objectives approved for entire Bay San Francisco Bay removed from 303(d) List in 2002
Dioxins / Furans	Updated assessment in 2018
Legacy Pesticides (Chlordane, Dieldrin, and DDT)	Monitoring recovery
Mercury	Bay TMDL and site-specific objectives approved in 2008 Guadalupe River Watershed TMDL approved in 2010
Bacteria	Richardson Bay TMDL adopted in 2008 Bay beaches (multiple listings); TMDL approved in 2017
PCBs	Bay TMDL approved in 2009
Selenium	North Bay TMDL approved in 2016
Trash	Municipalities required to implement trash load controls in 2009
Dissolved Oxygen	Site-specific objectives for Suisun Marsh approved in 2019

Cormorants on The Sisters ▶
(Amy Richey, SFEI)

BACWA. 2018. Group Annual Report: Nutrient Watershed Permit Annual Report. Bay Area Clean Water Agencies, Oakland, CA.

Baginska, B. 2015. Total Maximum Daily Load Selenium in North San Francisco Bay: Staff Report For Proposed Basin Plan Amendment. San Francisco Bay Regional Water Quality Control Board, Oakland, CA.

Chen, L., 2017. Water Column Selenium Concentrations in the San Francisco Bay-Delta: Recent Data and Recommendations for Future Monitoring. August 2017. SFEI Contribution #836.

Cloern JE., Jassby AD, Thompson JK, and Hieb KA. 2007. A cold phase of the East Pacific triggers new phytoplankton blooms in San Francisco Bay. PNAS 104: 18561-18565.

Cloern JE, Jassby DA. 2012. Drivers of change in estuarine-coastal ecosystems: Discoveries from four decades of study in San Francisco Bay. Rev Geophys 50(4). RG4001.

Custer CM, Custer TW, Dummer P, Etterson M, Thogmatin W, Wu Q, Kannan K, Trowbridge A and P McKann. 2013. Exposure and Effects of Perfluoroalkyl Substances in Tree Swallows Nesting in Minnesota and Wisconsin, USA. Arch Environ Contam Toxicol 66: 120-138.

Daum, T.; Lowe, S.; Toia, R.; Bartow, G.; Fairey, R.; Anderson, J.; Jones, J. 2000. Sediment Contamination in San Leandro Bay, CA. SFEI Contribution No. 48. San Francisco Estuary Institute: Oakland, CA.

Davis, J.A. 2017. The 25th Anniversary of the RMP. Pp. 8-21 in SFEI. 2017. The Pulse of the Bay: The 25th Anniversary of the RMP. SFEI Contribution #841. San Francisco Estuary Institute, Richmond, CA.

Davis, J.A., K. Abu Saba, and A.J. Gunther. 2001. Technical Report of the Sources, Pathways, and Loadings Workgroup. San Francisco Estuary Institute, Richmond, CA.

Davis, J.A., D. Yee, R. Fairey, and M. Sigala. 2017. San Leandro Bay Priority Margin Unit Study: Phase Two Data Report. San Francisco Estuary Institute, Richmond, CA. SFEI Contribution #855.

De La Cruz, S., Woo, I., Flanagan, A., and Mittelstaedt, H. 2017. Assessing the impact of periodic dredging on macroinvertebrate-prey availability for benthic foraging fishes: final study plan and preliminary pilot study results. SFEI Contribution No. 833. U.S. Geological Survey, Vallejo, CA.

Delaney, M., Goeden, B., and Goldbeck, S. 2008. Dredging and sediment management in San Francsico Bay. In The Pulse of the Estuary: monitoring and management water quality in the San Francisco Estuary. SFEI Contribution No. 559. San Francisco Estuary Institute, Richmond, CA.

EBMUD. 2016. Sewer System Management Plan. East Bay Municipal Utility District, Oakland, CA.

Gilbreath, A.N., Hunt, J.A., and McKee, L.J., 2019. Pollutants of Concern Reconnaissance Monitoring Progress Report, Water Years 2015-2018. Contribution No. 942. San Francisco Estuary Institute, Richmond, California.

Goals Project. 2015. The Baylands and Climate Change: What We Can Do. Baylands Ecosystem Habitat Goals Science Update 2015 prepared by the San Francisco Bay Area Wetlands Ecosystem Goals Project. California State Coastal Conservancy, Oakland, CA, 266 pp.

Green, M. and P.J. Shuler. 2019. MAP: The Bay Area Leads California in Population Growth. KQED News, April 2019.

Grieb, T., S. Roy, J. Rath, R. Stewart, J. Sun, and J.A. Davis. 2018. North Bay Selenium Monitoring Design. San Francisco Estuary Institute, Richmond, CA. SFEI Contribution #921.

Heal the Bay. 2019. 2018-19 Annual Beach Report Card. Heal the Bay, Santa Monica, CA

Heberger, M., R. Sutton, N. Buzby, J. Sun, D. Lin, M. Hladik, J. Orlando, C. Sanders, and E.T. Furlong. 2019. Current-Use Pesticides and Wastewater Contaminants in San Francisco Bay Margin Sediment and Water. SFEI Contribution No. 934. San Francisco Estuary Institute, Richmond, CA.

Lin, D., and Davis, J. 2018. Support for sediment bioaccumulation evaluation: toxicity reference values for the San Francisco Bay. SFEI Contribution No. 916. San Francisco Estuary Institute, Richmond, CA.

Linares-Casenave J, Linville, R, Van Eenennaam JP, Muguet JB, Doroshov SI. 2015. Selenium Tissue Burden Compartmentalization in Resident White Sturgeon (Acipenser transmontanus) of the San Francisco Bay Delta Estuary. Environmental Toxicology and Chemistry 34(1):152-160.

LTMS. 2001. Long-term management strategy for the placement of dredged material in the San Francisco Bay region. Prepared by the US Army Corps of Engineers, US Environmental Protection Agency, Bay Conservation and Development Commission, and San Francisco Bay Regional Water Quality Control Board. San Francisco, CA, 400 pp.

LTMS. 2013. Long-term management strategy for the placement of dredged material in the San Francisco Bay region, 12-year review final report. Prepared by the US Army Corps of Engineers, US Environmental Protection Agency, Bay Conservation and Development Commission, and San Francisco Bay Regional Water Quality Control Board. San Francisco, CA, 44 pp.

LTMS. 2018. Long-Term Management Strategy for the Placement of Dredged Sediment in the San Francisco Bay Region: Beneficial Reuse Fact Sheet. May 2018.

McKee, L.J., Gilbreath, A.N., Pearce, S.A. and Shimabuku, I., 2018. Guadalupe River mercury concentrations and loads during the large rare January 2017 storm. Contribution No. 837. San Francisco Estuary Institute, Richmond, California.

REFERENCES

Melwani, A.R., Yee, D., McKee, L., Gilbreath, A., Trowbridge, P., and Davis, J.A., 2018. Statistical Methods Development and Sampling Design Optimization to Support Trends Analysis for Loads of Polychlorinated Biphenyls from the Guadalupe River in San Jose, California, USA, Final Report.

MTC and ABAG. 2017. Plan Bay Area 2040. Metropolitan Transportation Commission and Association of Bay Area Governments. http://2040.planbayarea.org/reports

Novick, E., J. Wu, and D. Senn. 2015. Nutrients in Lower South Bay. Chapter 2 in Crauder,J., Downing-Kunz, M.A., Hobbs, J.A., Manning, A.J., Novick, E., Parchaseo, F., Wu, J., Schoellhamer, D.H., Senn, D.B., Shellenbarger, G.G., Thompson, J. and Yee, D. (2016). Lower South Bay Nutrient Synthesis. San Francisco Estuary Institute & Aquatic Science Center, Richmond. CA. Contribution # 732.

Perry, H., Lydon, A., Soumoy, P., and Goeden, B. 2015. San Francisco Bay Sediment: Challenges and Opportunities. Poster presented by BCDC at the 2015 State of the Estuary Conference, Oakland, CA.

Ross, B. D. 2012. Summary and evaluation of bioaccumulation test for total mercury conducted by San Francisco Bay dredging projects. LTMS Program, San Francisco, CA, 17 pp.

Sadaria AM, Sutton R, Moran KD, Teerlink J, Brown JV, Halden RU. 2017. Passage of fiproles and imidacloprid from urban pest control uses through wastewater treatment plants in northern California, USA. Environ Toxicol Chem DOI: 10.1002/etc3673.

Schoellhamer, D., McKee, L., Pearce, S., Kauhanen, P., Salomon, M., Dusterhoff, S., Grenier, L., Marineau, M., and Trowbridge P., 2018. Sediment Supply to San Francisco Bay, Water Years 1995 through 2016: Data, trends, and monitoring recommendations to support decisions about water quality, tidal wetlands, and resilience to sea level rise. SFEI Contribution Number 842. San Francisco Estuary Institute, Richmond, CA.

Schraga, T.S. and Cloern, J.E., 2017. Water quality measurements in San Francisco Bay by the US Geological Survey, 1969–2015. Scientific data, 4, p.170098.

SFBRWQCB. 2000. Beneficial reuse of dredged materials: sediment screening and testing guidelines. San Francisco Bay Regional Water Quality Control Board, Oakland, CA, 35 pp.

SFBRWQCB. 2019. San Francisco Bay Nutrients Watershed Permit, Revised Tentative Order No. R2-2019-0017. San Francisco Bay Regional Water Quality Control Board, Oakland, CA.

SFEI. 2017. Nutrient Management Strategy Science Program Annual Report. SFEI Contribution #838. San Francisco Estuary Institute, Richmond, CA.

Shimabuku, I., R. Sutton, D. Chen, Y. Wu, and J. Sun. 2019. Flame retardants and plastic additives in San Francisco Bay: Targeted monitoring of organophosphate esters and bisphenols. SFEI Contribution No. 925. San Francisco Estuary Institute, Richmond, CA.

Stewart AR, Luoma S, Schlekat C, Doblin M, and Hieb K. 2004. Food web pathway determines how selenium affects aquatic ecosystems: a San Francisco Bay case study. Environ. Sci. Technol. 38. 4519-4526.

Sun, J., J.A. Davis, A.R. Stewart, and V. Palace. 2019a. Selenium in White Sturgeon from North San Francisco Bay: The 2015-2017 Sturgeon Derby Study. SFEI Contribution #897. San Francisco Estuary Institute, Richmond, CA.

Sun, J., J.A. Davis, and A.R. Stewart. 2019b. Selenium in Muscle Plugs of White Sturgeon from North San Francisco Bay, 2015-2017. SFEI Contribution #929. San Francisco Estuary Institute, Richmond, CA.

Sutton, R., Y. Xie, K.D. Moran, and J. Teerlink. 2019a. Occurrence and Sources of Pesticides to Urban Wastewater and the Environment. In Goh, K.S., Gan, J., Young, D.F. and Luo, Y. eds., 2019. *Pesticides in Surface Water: Monitoring, Modeling, Risk Assessment, and Management*. American Chemical Society.

Sutton, R., D. Lin, M. Sedlak, C. Box, A. Gilbreath, A. Franz, R. Holleman, E. Miller, A. Wong, K. Munno, C. Rochman, and X. Zhu. 2019b. Understanding Microplastic Levels, Pathways, and Transport in the San Francisco Bay Region. SFEI Contribution #950. San Francisco Estuary Institute, Richmond, CA.

Tetra Tech, 2012. North San Francisco Bay Selenium Characterization Study. Final Report. http://www.waterboards.ca.gov/sanfranciscobay/water_issues/programs/TMDLs/northsfbayselenium/DSM2%20Selenium%20Loads%20from%20the%20Delta%205-1-2014.pdf

Tetra Tech, 2015. Updates to ECoS3 to simulate selenium fate and transport in North San Francisco Bay. Prepared for: San Francisco Bay Regional Water Resources Control Board.

Urquhart et al. 1991. Selenium Verification Study, 1988-1990. California State Water Resources Control Board, Sacramento, CA.

USEPA and USACE. 1991. Evaluation of dredged material proposed for ocean disposal – Testing Manual. EPA 503/8-91/001, Washington, D.C., 214 pp.

USEPA and USACE. 1998. Evaluation of dredged material proposed for discharge in waters of the U.S. – Testing Manual. EPA-823-B-98-004, Washington, D.C., 176 pp.

Yee, D.; Bemis, B.; Hammond, D.; Rattonetti, T.; van Bergen, S. 2011. Age Estimates and Pollutant Concentrations of Sediment Cores from San Francisco Bay and Wetlands. SFEI Contribution 652. San Francisco Estuary Institute: Richmond, CA.

Yee, D., Trowbridge, P., and Sun, J. 2015. Updated ambient concentrations of toxic chemicals in San Francisco Bay sediments. SFEI Contribution No. 749. San Francisco Estuary Institute, Richmond, CA.

Yee, D., and Wong, A. 2019. Evaluation of PCB concentrations, masses, and movement from dredged areas in San Francisco Bay. SFEI Contribution No. 938. San Francisco Estuary Institute, Richmond, CA.

Yee, D., Wong, A. and Hetzel, F. 2019. Current knowledge and data needs for dioxins in San Francisco Bay. SFEI Contribution No. 926. San Francisco Estuary Institute, Richmond, CA.

GRAPH DETAILS

Page 70

DELTA SEDIMENT LOAD
Loads based on continuous measurements taken at Mallard Island by USGS (http://sfbay. wr.usgs.gov/sediment/cont_ monitoring/). Data are for water years (October 1 to September 30 with the year corresponding to the end date)

SUSPENDED SEDIMENT
Data for Dumbarton Bridge, 20 feet below mean lower low water. Based on 15-minute data collected by the U.S. Geological Survey (Buchanan et al. 2014). Data gap during WY2012 and 2013 due to construction for seismic retrofit of highway bridge.

Page 72

SEA LEVEL
Data from National Oceanic and Atmospheric Administration. Data and more information available at: https://tidesandcurrents.noaa. gov/sltrends/sltrends_station. shtml?stnid=9414290

Page 73

CHLOROPHYLL
After Cloern et al. (2007) and SFEI (2017). Based on near-surface (0-2 m) data from August-December using the same stations (s21,s22,s24,s25,s27,s29,s30,s32) and following the same averaging approach as described in Cloern et al. 2007. Data collected monthly at fixed stations along the spine of the Bay. Data are from the USGS water quality component of the RMP and available online (Schraga and Cloern 2017).

Page 75

IMPORTANT NUTRIENT PARAMETERS
Data are from the USGS water quality component of the RMP and available online (Schraga and Cloern 2017).

Page 84

GUADALUPE RIVER FLOW
Data from the US Geological Survey. Data for are for water years (Oct 1 to Sep 30).

GUADALUPE RIVER MERCURY LOAD
Total loads for each water year (Oct 1 to Sep 30). Additional matching funds for this study provided by the CEP, USACE, SCVWD, and SCVURPPP. Data from McKee et al. (2018) and related publications.

◀ An oil/chemical tanker off Hunters Point
(Amy Richey, SFEI)

RMP COMMITTEE MEMBERS AND PARTICIPANTS

RMP Steering Committee

BACWA Principle, Eric Dunlavey, City of San Jose

BACWA Associate, Leah Godsey Walker, City of Petaluma

BACWA Associate, Karin North, City of Palo Alto

Stormwater Agencies, Adam Olivieri, EOA, Inc.

Dredgers, John Coleman, Bay Planning Coalition

Dredgers, Maureen Dunn, Chevron

San Francisco Bay Regional Water Quality Control Board, **Tom Mumley**

US Army Corps of Engineers, Tawny Tran

RMP Steering Committee Chair in bold

RMP Technical Review Committee

BACWA, Mary Lou Esparza, Central Contra Costa Sanitary District

BACWA, Yuyun Shang, East Bay Municipal Utility District

South Bay Dischargers, Tom Hall, EOA, Inc.

City and County of San Francisco, Ross Duggan,

City of San Jose, Anne Hansen,

Refineries, **Bridgette DeShields,** Integral Consulting

Stormwater, Chris Sommers, EOA, Inc.

Dredgers, Shannon Alford, Port of San Francisco

San Francisco Bay Regional Water Quality Control Board, Richard Looker

USEPA Region IX, Luisa Valiela

US Army Corps of Engineers, Jim Mazza

NGO, Ian Wren, Baykeeper

RMP Technical Review Committee Chair in bold

RMP Science Advisors

EMERGING CONTAMINANTS WORKGROUP

Dr. Bill Arnold, University of Minnesota

Dr. Lee Ferguson, Duke University

Dr. Derek Muir, Environment Canada

Dr. Kelly Moran, TDC Environmental, LLC

Dr. Heather Stapleton, Duke University

Dr. Miriam Diamond, University of Toronto

MICROPLASTICS WORKGROUP

Dr. Anna-Marie Cook, USEPA Region IX

Dr. Chelsea Rochman, University of Toronto

Dr. Kara Lavender Law, Sea Education Association

PCB WORKGROUP

Dr. Frank Gobas, Simon Fraser University

SEDIMENT WORKGROUP

Dr. David Schoellhamer, USGS Emeritus

Dr. Patricia Wiberg, University of Virginia

SELENIUM WORKGROUP

Dr. Harry Ohlendorf, Independent

SOURCES, PATHWAYS, AND LOADINGS WORKGROUP

Dan Cain, USGS

Barbara Mahler, USGS

Tom Jobes, Independent

Dan Wang, CA Department of Pesticide Regulation

RMP Participants

MUNICIPAL DISCHARGERS

City of Benicia

City of Burlingame

City of Calistoga

Central Contra Costa Sanitary District

Central Marin Sanitation Agency

Delta Diablo

East Bay Dischargers Authority

East Bay Municipal Utility District

Fairfield-Suisun Sewer District

Las Gallinas Valley Sanitary District

City of Millbrae

Mountain View Sanitary District

Napa Sanitation District

Novato Sanitation District

City of Palo Alto

City of Petaluma

City of Pinole/Hercules

Rodeo Sanitary District

San Francisco International Airport

City and County of San Francisco

City of San Jose

City of San Mateo

Sausalito-Marin City Sanitary District

Sewerage Agency of Southern Marin

City of South San Francisco/San Bruno

Sonoma County Water Agency

Silicon Valley Clean Water

City of Sunnyvale

City of St. Helena

Marin County Sanitary District #5, Tiburon

Union Sanitary District

Vallejo Flood and Wastewater District

West County Wastewater District

Town of Yountville

U.S. Navy, Treasure Island

INDUSTRIAL DISCHARGERS

C&H Sugar Company

Chevron Products Company

Crockett Cogeneration

Eco Services Operations Corporation

Phillips 66 Company

Shell Martinez Refinery

Tesoro Golden Eagle Refinery

USS-POSCO Industries

Valero Refining Company

STORMWATER

Alameda County Clean Water Program

California Department of Transportation

City and County of San Francisco

Contra Costa Clean Water Program

Fairfield-Suisun Urban Runoff Management Program

Marin County Stormwater Pollution Prevention Program

Santa Clara Valley Urban Runoff Pollution Prevention Program

San Mateo Countywide Water Pollution Prevention Program

Vallejo Sanitation & Flood Control District

DREDGERS

Belvedere Cove Access Channel

Benicia Port Terminal Company Pier 95

Chevron Richmond Long Wharf

City of Benicia Marina

Glen Cove Marina

Marina Dredge Neighbors

Phillips 66 Company Rodeo Terminal

Port of Oakland

Port of Redwood City

Port of San Francisco

San Francisco Yacht Club

U.S. Coast Guard Environmental Division

Vallejo Yacht Club

◀ **A Microplastics Workgroup meeting**
(Jay Davis, SFEI)

CREDITS AND ACKNOWLEDGEMENTS

Editors

Jay Davis

Melissa Foley

Contributing Authors

Jay Davis

RMP Data Management

Cristina Grosso

Amy Franz

John Ross

Don Yee

Adam Wong

Information Compilation

Amy Franz

John Ross

Adam Wong

Lester McKee

Dave Senn

Cristina Grosso

Shira Bezalel

Richard Looker

Randy Turner

Noushin Adabi

Ila Shimabuku

Design and Production

Ruth Askevold

Amy Richey

Katie McKnight

Information Graphics

Linda Wanczyk (lindawanczyk.com)

Amy Richey

Adam Wong

Katie McKnight

Micaela Bazo

Alicia Gilbreath

Maureen Downing-Kunz

Ellen Plane

The following reviewers greatly improved this document by providing comments on draft versions:

Dan Glaze

Leah Walker

Tom Hall

Jim Cloern

Dave Senn

Richard Looker

Bill Johnson

Kelly Moran

Ian Wren

Eric Dunlavey

Luisa Valiela

Dave Halsing

Barbara Baginska

Harry Ohlendorf

Bonnie de Berry

Maureen Dunn

Maureen Downing-Kunz

Marisol Pacheco-Mendez

James Downing

Jon Konnan

Lucile Paquette

Stu Townsley

Ila Shimabuku

Meg Sedlak

Don Yee

Rebecca Sutton

Lester McKee

Scott Dusterhoff

Warner Chabot

▲ A Brown Pelican rehabilitated by International Bird Rescue (bird-rescue.org) being released at Fort Baker (Cheryl Reynolds, International Bird Rescue)

RMP

REGIONAL MONITORING PROGRAM FOR WATER QUALITY IN SAN FRANCISCO BAY
sfei.org/rmp

Administered by the San Francisco Estuary Institute

4911 Central Avenue, Richmond, CA 94804
p: 510-746-SFEI (7334), f: 510-746-7300

www.sfei.org

www.ingramcontent.com/pod-product-compliance
Lightning Source LLC
Chambersburg PA
CBHW041100210326

41597CB00005B/144